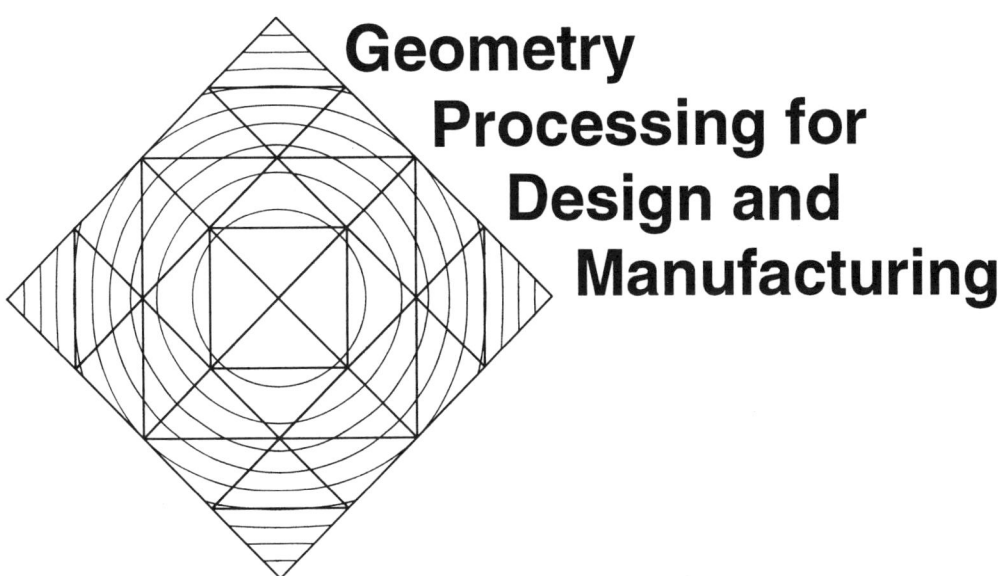

Geometry Processing for Design and Manufacturing

Geometry Processing for Design and Manufacturing

Edited by
Robert E. Barnhill
Arizona State University

siam.

Philadelphia

Society for Industrial and Applied Mathematics

Geometric Design Publications

Editor
Gerald E. Farin
Arizona State University

Published Volumes

Farin, Gerald E., editor, Geometric Modeling: Algorithms and New Trends (1987)
Farin, Gerald E., editor, NURBS for Curve and Surface Design (1991)
Barnhill, Robert E., editor, Geometry Processing for Design and Manufacturing (1992)

Volumes in Preparation

Hagen, Hans, editor, Curve and Surface Design (1992)
Hagen, Hans, editor, Topics in Surface Modeling (1992)

Library of Congress Cataloging-in-Publication Data

Geometry processing for design and manufacturing \ edited by Robert E.
    Barnhill
        p.   cm.
    Outgrowth of talks at the Geometric Design Conference of the
Society for Industrial and Applied Mathematics, Tempe, Arizona,
November 1989.
    Includes bibliographical references and index.
    ISBN 0-89871-280-7
    1. Geometry--Data processing--Congresses.   2. Engineering design-
-Data processing--Congresses.   3. Design, Industrial--Data
processing--Congresses.   I. Barnhill, Robert E.   II. Society for
Industrial and Applied Mathematics.   III. Geometric Design
Conference (1989 : Tempe, Ariz.)
QA448.D38G46   1992   91-46596
516'.0285--dc20

Sponsored by SIAM Activity Group on Geometric Design.

Preface

Geometry processing is the calculation of geometric properties of already constructed curves, surfaces, and solids. In computer aided geometric design and related subjects, much attention has been devoted to the design of curves, surfaces, and solids, but considerably less attention has been given to determining their geometric properties. Geometry processing methods must be employed in any functional CAD/CAM system. But since geometry processing is intrinsically "hard," no elegant, unified theory has been developed thus far. This volume is an attempt to spur such a unifying development.

The geometry processing topics herein include curvature analysis, curve fairing, contouring, conversions between curves and between surfaces, intersections between surfaces, and offset curves and surfaces. The calculation of offset curves and surfaces and of the intersections between surfaces are fundamental problems in geometry processing. These two topics form the foundation of this book.

Accordingly, the papers are organized into two groups, under the headings of offset curves and surfaces and surface-surface intersections (SSI), respectively. More precisely, the first five papers concern aspects of offsets and the last four, plus a bibliography, involve intersections. Within the two groups, the papers are ordered to link subtopics as smoothly as possible.

Although the papers tell their own stories, we add a few words here about each paper to aid the reader.

Chapter 1. Offset curves and surfaces are difficult in part because they do not belong to the same function classes as their progenitors. Farouki develops special offset curves which can be parameterized by rational functions and thus would fit easily into current modeling systems.

Chapter 2. A major difficulty in the computation of offset surfaces is their self-intersections: this topic is addressed by Barnhill, Frost, and Kersey, who use a combination of geometric and numerical techniques to find self-intersections of networks of general triangular or rectangular parametric patches. The generality is important in that the algorithms are not restricted to networks of polynomial or rational patches, so that the offset surfaces need not be approximated.

Chapter 3. Communication between systems using different curve forms or between different surface forms is a fundamental problem of considerable practical importance. An intrinsic problem in such portability is the (approximate) conversion between different curve and surface forms. Hoschek and Schneider develop suitable approximations for conversions between B-spline surfaces, which are also applied to offsets.

Chapter 4. Farin presents and compares two classes of methods for the fairing of B-spline curves and surfaces, "knot removal fairing" and "degree reduction fairing." Curvature analysis, an additional topic in geometry processing, is an essential tool in this research.

Chapter 5. Brechner presents general, "envelope" methods for determining offset curves and surfaces. These general results permit, as corollaries, a variety of examples and applications including tool path generation.

Chapter 6. Surfaces are frequently visualized and/or interrogated by means of their contours. Moreover, in the spirit of this volume, contours of a surface may be considered as a first example of surface-surface intersection. Barnhill, Bloomquist, and Worsey develop adaptive contouring algorithms for the contouring of surfaces that are networks of triangular polynomial patches.

Chapter 7. Many mechanical objects are constructed from surfaces that have special mathematical forms, such as natural quadrics (plane, cone, cylinder, and sphere) and extruded surfaces. Algorithms for these often-occurring cases are illustrated in the paper by Piegl.

Chapter 8. Patrikalakis develops surface-surface intersection algorithms for two classes of problems: implicit-parametric and parametric-parametric intersections, respectively. The implicit surfaces are polynomial and the parametric surfaces are rational. A key ingredient is the calculation of significant points (border, turning, and singular points) of the surfaces, because intersections are monotonic within the corresponding subpatches.

Chapter 9. Wang addresses the intersection problem for rational parametric surfaces by examining the intersection between isoparametric curves and the second surface. A distance evaluation function is introduced to improve the detection and calculation of intersections.

Chapter 10. Because the subject of calculating surface-surface intersections is developing so rapidly, Farin has prepared a modern bibliography on SSI.

This book was motivated by the many talks on geometry processing at the Geometric Design Conference of the Society for Industrial and Applied Mathematics held in Tempe, Arizona in November 1989. The considerable interest in geometry processing at that conference was shown by the large number of talks and discussions on the topic. Geometry processing has many applications: in CAD/CAM/CIM via solid modeling and numerically controlled machining as well as in scientific computing via contouring. These and other applications account for some of its considerable popularity. However, geometry processing also has interest as a scientific subject in its own right. The papers presented herein illustrate both of these attractions of the subject.

This book represents a team effort of many people. The papers in this volume were specifically requested, to represent a wide yet coherent cross-section of current geometry processing. The splendid cooperation of the selected authors shows in the high quality of their papers. The referees of the manuscripts deserve both thanks and public recognition: Gerald Farin, Rida Farouki, Todd Frost, Dianne Hansford, Scott Kersey, Eugene Lee, Robert Magedson, Robert Markot, Ray Sarraga, and Tom Sederberg. Thanks for assistance in the preparation of the manuscripts also go to Carol Chapman, Joe Reuter, and Wayne Woodland in the Computer Science Department at Arizona State University. The Arizona State University portion of this work was supported by the Department of Energy under grant DE-FG02-87ER25041 and by the National Science Foundation under grant DMC-8807747.

Robert E. Barnhill
Arizona State University

List of Contributors

Robert E. Barnhill, Department of Computer Science and Engineering, Arizona State University, Tempe, AZ 85287-5406

Brett K. Bloomquist, Spatial Technology, 2425 55th Street, Boulder, CO 80301

Eric L. Brechner, Boeing Computer Services, P. O. Box 24346, Mail Stop 7L-24, Seattle, WA 98124-0346

Gerald Farin, Department of Computer Science and Engineering, Arizona State University, Tempe, AZ 85287-5406

Rida T. Farouki, IBM Thomas J. Watson Research Center, P.O. Box 218, Yorktown Heights, NY 10598

Todd M. Frost, Department of Computer Science and Engineering, Arizona State University, Tempe, AZ 85287-5406

Josef Hoschek, Fachbereich Mathematik, Technische Hochschule, Darmstadt, D-6100 Darmstadt, Germany

Scott N. Kersey, Department of Mathematics, University of Wisconsin, Madison, WI 53706

Nicholas M. Patrikalakis, Department of Ocean Engineering, Design Laboratory, MIT, Cambridge, MA 02139

Les A. Piegl, Computer Science Department, University of Southern Florida, Tampa, FL 33620-5399

Franz-Josef Schneider, Fachbereich Mathematik, Technische Hochschule, Darmstadt, D-6100 Darmstadt, Germany

K. Y. Wang, General Electric Company, GE Corporate Research & Development, P.O. Box 8, Schenectady, NY 12301

Andrew J. Worsey, General Electric Company, GE Corporate Research & Development, P.O. Box 8, Schenectady, NY 12301

Contents

Offset Curves and Surfaces

Pythagorean–Hodograph Curves in Practical Use

Rida T. Farouki

1.1. Introduction

In comparing the problems of *designing* curves to satisfy various interpolatory, smoothness, and aesthetic requirements with those of *processing* curves in order to derive certain "procedurally defined" geometric forms from them, one is struck by a marked imbalance in the degree of attention these problems have received from the computer-aided geometric design (CAGD) community, relative to the mathematical difficulties they entail and their practical importance.

Among the most familiar of procedurally defined curves are the space curves generated by the intersections of algebraic surfaces, which play a key role in solid modeling, and the "offsets" to plane polynomial parametric curves, which arise in applications such as tolerance analysis and numerical-control machining. Unfortunately, both of these forms are fundamentally incompatible with the simple (rational) parametric representations that are the accepted paradigm for design problems.

It is the author's perception that while the plethora of different curves propounded for design purposes has surpassed the engineering requirements that originally motivated it, the numerical schemes available thus far for dealing with offsets and intersections fail to meet the minimum standards of reliability, accuracy, and efficiency that geometric modeling systems must aspire to if they are to serve their intended role in a computer-integrated manufacturing environment.

This paper is an informal survey of recent attempts to address the dilemma described above (in the case of offset curves, at least). Specifically, we shall describe a special class of polynomial parametric curves whose offsets admit *precise* parameterizations by rational functions, and are thus fully compatible with the representational and algorithmic conventions of contemporary modeling systems.

Let us remind the reader that for any regular plane parametric curve $\mathbf{r}(t) = \{x(t), y(t)\}$, an *offset* $\mathbf{r}_o(t)$ is a locus at a fixed (signed) distance d

from it, measured in the direction of its unit normal vector $\mathbf{n}(t)$:

$$(1.1) \qquad\qquad \mathbf{r}_o(t) = \mathbf{r}(t) + d\,\mathbf{n}(t)\,.$$

Notwithstanding the conceptual simplicity of this procedural definition for an offset curve, such loci exhibit a remarkably rich geometric and algebraic structure (see [14], [15] for a thorough description).

An idea of the importance of offset curves in practical applications can be gained from the great volume of literature devoted to schemes for approximating them in terms of simple (piecewise polynomial or rational) functional forms (see, for example, [30], [41], [26]–[28], [5], [34], [10]). Most of these approximation schemes rely upon rather naïve sampling and subdivision strategies; they generally make no explicit a priori analysis of the offset topology and singularities, and are consequently prone to failure under extreme circumstances.

1.2. Hodographs

In order to motivate the ensuing discussion, we need to make a slight detour at this point in order to review the notion of a hodograph. Although it has occasionally been invoked in the CAGD literature by previous authors (e.g., [1], [37], [36], [9]), it seems that the term "hodograph" is still widely regarded as somewhat arcane.

The *hodograph* of a plane parametric curve $\mathbf{r}(t) = \{x(t), y(t)\}$ is simply the locus described by the first parametric derivative $\mathbf{r}'(t) = \{x'(t), y'(t)\}$ of that curve. Hodographs are actually a well-established concept of classical mechanics (where the curve $\mathbf{r}(t)$ is imagined as describing the trajectory of a particle, the parameter t representing time, and the derivative $\mathbf{r}'(t)$ the particle velocity [40]), but they have recently lapsed into obscurity. To quote from *Classical Mechanics* [20]:

> "If the velocity vector of a particle is translated so as to start from the center of force, then the head of the vectors traces out the particle's *hodograph*, a locus of considerable antiquity in the history of mechanics ..."

Figure 1.1 illustrates this idea (note that we introduce the analogy with particle motion only for visualization purposes; any differentiable parametric curve exhibits a continuous hodograph).

Let us focus henceforth on the hodographs of plane curves described by *polynomial* parameterizations. If $\mathbf{r}(t) = \{x(t), y(t)\}$ is a polynomial curve of degree n,

$$(1.2) \qquad\qquad x(t) = \sum_{k=0}^{n} a_k t^k\,, \qquad y(t) = \sum_{k=0}^{n} b_k t^k\,,$$

then its hodograph $\mathbf{r}'(t)$ is evidently a polynomial curve of degree $n - 1$. For each t, the *magnitude* or "length" of the hodograph is given by

$$(1.3) \qquad\qquad \sigma(t) = |\mathbf{r}'(t)| = \sqrt{x'^2(t) + y'^2(t)}\,,$$

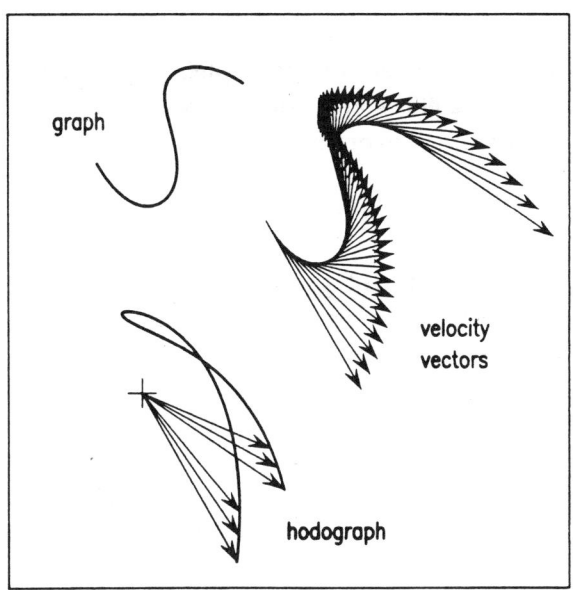

FIG. 1.1. *An S-shaped parametric curve and its hodograph.*

and if s denotes the arc length measured along $\mathbf{r}(t)$ from some fixed point, we observe that

$$(1.4) \qquad \sigma(t) = \frac{ds}{dt}.$$

For this reason, we shall also refer to the hodograph magnitude (1.3) as the *parametric speed* of the curve.

Now from a computational standpoint, the polynomial curves (1.2) are perhaps the simplest imaginable class of curves for practical use, but the functional form (1.3) of their parametric speed $\sigma(t)$ is nevertheless the source of two rather annoying facts about these curves. First, if we need to know the cumulative arc-length $s(t)$ of such a curve (measured from $t = 0$), it is given by the recalcitrant integral

$$(1.5) \qquad s(t) = \int_0^t \sqrt{x'^2(\tau) + y'^2(\tau)}\, d\tau,$$

which in general yields only to a brute-force numerical approximation, for each value of t. (Notable exceptions are the cases $n = 1$ and $n = 2$, which correspond to linear and parabolic arcs.)

Second, while the normal line at each point of the curve (1.2) has the direction $(y'(t), -x'(t))$, in order to measure a prescribed distance d along this normal line we need to form the *unit* normal vector

$$(1.6) \qquad \mathbf{n}(t) = \frac{(y'(t), -x'(t))}{\sqrt{x'^2(t) + y'^2(t)}}.$$

Obviously $\mathbf{n}(t)$ does *not* (in general) depend rationally on the parameter t, and the offset curves (1.1) consequently lie outside the usual representational conventions of CAGD.

At this point the reader may be excused for wondering how Pythagoras could be implicated in the matter of hodographs, since he anteceded the notion of a derivative by some 2000 years. Indeed, we associate his name almost exclusively with the famous "theorem"

$$(1.7) \qquad\qquad\qquad a^2 + b^2 = c^2,$$

relating the hypotenuse length c of a right-angle triangle to the lengths a and b of the other sides. Let us therefore review (to borrow a phrase from a well-known opera score by W. S. Gilbert) some "cheerful facts about the square on the hypotenuse."

Usually, we think of (a, b, c) in (1.7) as positive real numbers, in which case we are free to choose a and b at will, with (1.7) then furnishing the corresponding positive real value of c. For our present purposes, however, it is more interesting to note that (1.7) can only be satisfied in exceptional cases — the "Pythagorean triples" — when (a, b, c) are *integers*.

The oldest and simplest (nontrivial) Pythagorean triple is $(3, 4, 5)$. The engineers charged with the construction of the pyramids in Egypt more than 4000 years ago recognized that a triangle whose sides were 3, 4, and 5 units in length always formed a perfect right angle or "set square." The Babylonians knew many more Pythagorean triples, such as $(8,15,17)$ and even $(3367, 3456, 4825)$; this remarkable Mesopotamian numeracy may be attributable to the fact that they used sexagesimal (base 60) arithmetic.

However, it was not until about 550 B.C. that Pythagoras proved that the sides of *any* right-angle triangle must satisfy the famous equation (1.7). This was arguably the first great quantitative law of science, offering a nonobvious, systematic, and accurate predictive capacity [2].

To Pythagoras and his predecessors the *integer* triples satisfying (1.7) were evidently somewhat magical, but today it is an established fact of elementary number theory [29] that the totality of these solutions has a quite simple characterization: an integer triple (a, b, c) satisfies (1.7) *if and only if* it can be expressed in terms of another integer triple (u, v, w) in the form

$$(1.8) \qquad a = w(u^2 - v^2), \quad b = 2wuv, \quad c = w(u^2 + v^2)$$

(obviously, the roles of a and b are interchangeable). For example, with the choice $(u, v, w) = (2, 1, 1)$, we obtain the familiar $(a, b, c) = (3, 4, 5)$.

While we were engrossed in a somewhat different problem (see [19]) in the spring of 1989, Takis Sakkalis brought to my attention the notion of Pythagorean triples of (real) *polynomials*, and he unearthed a short paper [31] showing that such triples were amenable to exactly the same characterization (1.8) as the integers.

Although it was not immediately clear that this offered the key to solving our original problem — proving that it is impossible to parameterize any real plane curve by rational functions of its arc length — it occurred to me that hodographs defined by Pythagorean polynomial triples would be of interest in their own right, for the reasons described herein.

If $(a(t), b(t), c(t))$ are polynomials with real coefficients that may be expressed in terms of other real polynomials $(u(t), v(t), w(t))$ in the form

(1.9)
$$\begin{aligned} a(t) &= w(t)[u^2(t) - v^2(t)], \\ b(t) &= 2\, w(t)u(t)v(t), \\ c(t) &= w(t)[u^2(t) + v^2(t)], \end{aligned}$$

then it is clear that the condition

(1.10)
$$a^2(t) + b^2(t) = c^2(t)$$

is satisfied for each — real and complex — value of t. That *every* triple of real polynomials $(a(t), b(t), c(t))$ satisfying (1.10) for all t must be of the form (1.9) for *some* $(u(t), v(t), w(t))$ is a more subtle matter.

Actually, there is a simple Pythagorean triple of polynomials with which the reader is probably already acquainted. Consider the rational parameterization of the unit circle,

(1.11)
$$x(t) = \frac{X(t)}{W(t)}, \qquad y(t) = \frac{Y(t)}{W(t)},$$

where

(1.12)
$$X(t) = 1 - t^2, \quad Y(t) = 2t, \quad W(t) = 1 + t^2,$$

obtained by making the "half angle substitution" $t = \tan \frac{1}{2}\theta$ in the familiar trigonometric form $(x, y) = (\cos\theta, \sin\theta)$. Clearly $x^2(t) + y^2(t) \equiv 1$ implies that

(1.13)
$$X^2(t) + Y^2(t) \equiv W^2(t),$$

so we see that $(a(t), b(t), c(t)) = (1 - t^2, 2t, 1 + t^2)$ is a Pythagorean triple, corresponding to the choice $(u(t), v(t), w(t)) = (1, t, 1)$ in (1.9).

Suppose now that we assign the first two elements $1 - t^2$ and $2t$ of the Pythagorean triple (1.12) to be the *components of a hodograph*,

(1.14)
$$x'(t) = 1 - t^2, \qquad y'(t) = 2t.$$

On integration, we obtain from (1.14) the Pythagorean–hodograph curve

(1.15)
$$x(t) = \frac{1}{3}t(3 - t^2) + x_0, \qquad y(t) = t^2 + y_0.$$

Note that the parametric speed of this curve is the *polynomial* expression $\sigma(t) = 1 + t^2$ in the parameter t, rather than the radical form (1.3).

Figure 1.2 shows the hodograph (1.14) and corresponding Pythagorean–hodograph curve (1.15) with the choice $(x_0, y_0) = (0, 0)$. This curve has the remarkable property that its offsets are *rational* curves and its arc length is a *polynomial* function $s(t)$ of the parameter t. We will have occasion to return to it!

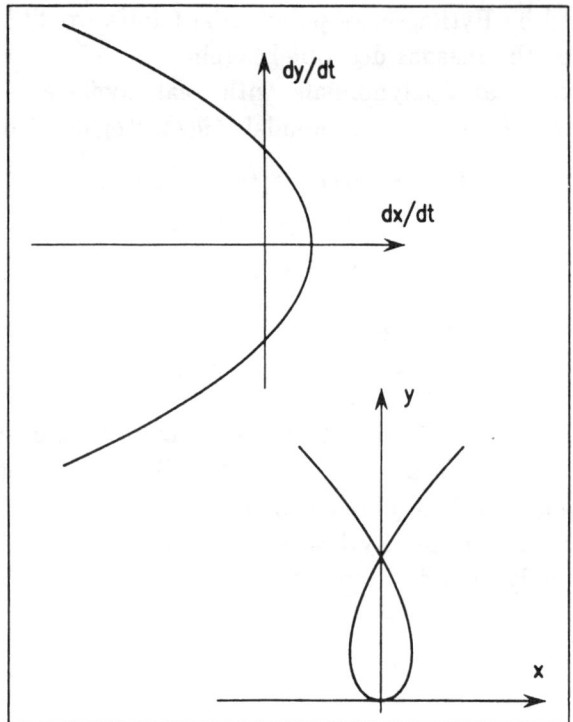

FIG. 1.2. *Integrating the hodograph* $(x', y') = (1 - t^2, 2t)$.

1.3. Pythagorean–Hodograph Curves

By the expression "Pythagorean–hodograph curve" we shall henceforth mean any plane polynomial parametric curve whose hodograph components $\{x'(t), y'(t)\}$ and magnitude $\sigma(t)$ are the elements of a (real) Pythagorean triple of polynomials of the form (1.9).

In particular, we shall focus primarily on the odd-degree curves defined by hodographs of the form

$$(1.16) \qquad x'(t) = u^2(t) - v^2(t), \qquad y'(t) = 2u(t)v(t)$$

for real, nonzero polynomials $u(t)$ and $v(t)$ that are not *both* constants and have no factors in common (the Pythagorean–hodograph curves with $w(t) \neq 1$ have certain undesirable qualities that are described in [18]).

On integrating the forms (1.16) to obtain a Pythagorean–hodograph curve $\mathbf{r}(t) = \{x(t), y(t)\}$, it is clear that $\mathbf{r}(t)$ will be a cubic when the polynomials $u(t)$ and $v(t)$ are linear, and a quintic when they are quadratic. The former proves to be perhaps more a curiosity than a practical design tool; it is the latter that hold promise for engineering use, since these special quintics are similar to regular cubics in their "shape freedoms."

The purpose of this article is to promote the transfer of Pythagorean–hodograph curves to practical use by describing in simple terms their basic characteristics and the advantages they offer over parametric polynomial curves in general (of which they form a subset). At the same time, we will be quite frank about the limitations of which we are currently aware.

Thus, we shall eschew detailed analyses and proofs of theorems (which have been sequestered elsewhere: [18], [13]), concentrating here instead on those aspects that are perceived to be of greatest importance in implementation and practical use. We shall also frequently illustrate key concepts by means of numerical examples.

We rely exclusively on the Bernstein–Bézier representation

$$(1.17) \qquad \mathbf{r}(t) = \sum_{k=0}^{n} \mathbf{p}_k \binom{n}{k} (1-t)^{n-k} t^k \quad \text{for } t \in [0,1]$$

of (finite) parametric polynomial curves, and we assume that the reader is familiar with the basic properties and terminology associated with this representation (if not, see [9] for a review). In particular, we shall seek constraints on the "control points" $\{\mathbf{p}_k\}$ and/or procedures for their construction, such that the curve (1.17) is guaranteed to have a Pythagorean hodograph.

One aspect of the form (1.17) that is especially important in our present context is its innate numerical stability [16]. Since the Pythagorean–hodograph curves form a lower dimension subset of the space of parametric polynomial curves of a given degree, any perturbation of the control points $\{\mathbf{p}_k\}$ defining such a curve will almost invariably destroy its Pythagorean–hodograph property. Thus, for floating point computations, we shall actually be dealing with curves whose hodographs are "nearly" Pythagorean.

The attractive features of Pythagorean–hodograph curves (polynomial arc length functions and rational offset curves) must therefore be regarded as only approximate in a floating point environment, and the stability of the polynomial basis in which we formulate our algorithms may exert a significant influence on the accuracy of computed results.

We will return to this matter again, presenting empirical evidence that, although one encounters higher degrees in dealing with Pythagorean–hodograph forms, excellent results can be obtained when using the Bézier representation in standard double precision floating point (the *arithmetic* errors incurred in computing the nominally exact offsets to Pythagorean–hodograph quintics are vastly smaller than the *approximation* errors arising in typical schemes approximating offsets by piecewise cubics).

We have already encountered a cubic Pythagorean–hodograph curve in §1.2 (see Fig. 1.2); let us now describe such cubics in greater detail. For $u(t)$ and $v(t)$ in (1.16) we choose the linear polynomials

$$(1.18) \qquad u(t) = u_0(1-t) + u_1 t, \qquad v(t) = v_0(1-t) + v_1 t,$$

expressed in Bernstein–Bézier form. Upon substituting (1.18) into (1.16) and integrating we infer that, for real values of (u_0, u_1) and (v_0, v_1), control points of the form

$$\mathbf{p}_1 = \mathbf{p}_0 + \frac{1}{3}(u_0^2 - v_0^2, 2u_0 v_0),$$

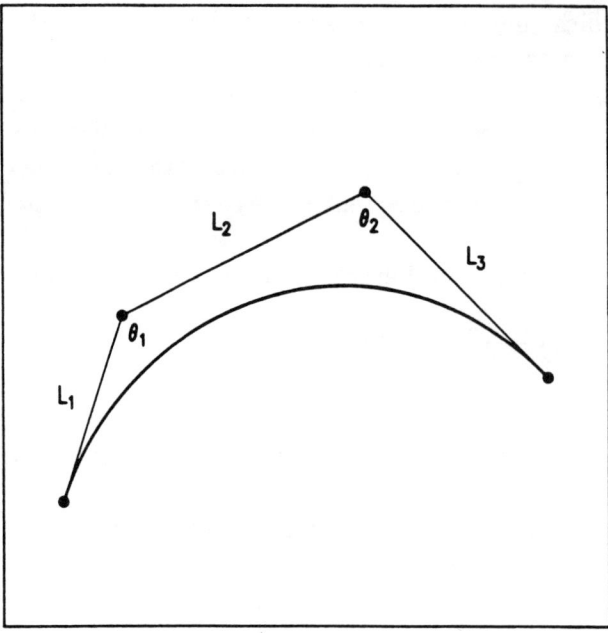

FIG. 1.3. *The Bézier control polygon of a cubic.*

$$(1.19) \qquad \mathbf{p}_2 = \mathbf{p}_1 + \frac{1}{3}(u_0 u_1 - v_0 v_1, u_0 v_1 + u_1 v_0),$$

$$\mathbf{p}_3 = \mathbf{p}_2 + \frac{1}{3}(u_1^2 - v_1^2, 2 u_1 v_1),$$

will define a cubic Pythagorean–hodograph curve (the first control point $\mathbf{p}_0$ being chosen at will).

There is a simple and elegant geometric characterization of the control polygon of a cubic that corresponds to a sufficient and necessary condition for a Pythagorean hodograph. Let L_1, L_2, L_3 be the lengths of the control polygon legs

$$(1.20) \qquad L_1 = |\mathbf{p}_1 - \mathbf{p}_0|, \quad L_2 = |\mathbf{p}_2 - \mathbf{p}_1|, \quad L_3 = |\mathbf{p}_3 - \mathbf{p}_2|,$$

and let θ_1 and θ_2 be the control polygon angles at the interior vertices $\mathbf{p}_1$ and $\mathbf{p}_2$ (see Fig. 1.3). Then the cubic has a Pythagorean hodograph if and only if [18]:

$$(1.21) \qquad L_2 = \sqrt{L_1 L_3} \quad \text{and} \quad \theta_1 = \theta_2,$$

i.e., the length of the middle leg is the *geometric mean* of the lengths of the outer legs, and the angles are equal. The constraints (1.21) offer a much more intuitive means of dealing with Pythagorean–hodograph cubics.

Now a rather detailed analysis [18] reveals the somewhat disappointing fact that — allowing for rigid motions, uniform scalings, and linear reparameterizations — there is just *one* Pythagorean–hodograph cubic! It corresponds to a classical curve, known variously as *Tschirnhausen's cubic*, *l'Hôpital's cubic*, or

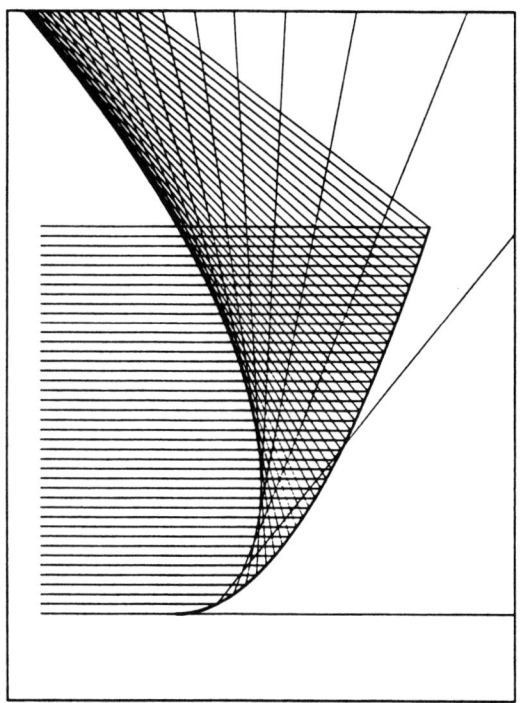

FIG. 1.4.   *Parallel light rays reflected by a parabola.*

the *trisectrix of Catalan* (see [32] or the three recently reprinted volumes of *Traité des Courbes Spéciales Remarquables Planes et Gauches* [21]).

Tschirnhausen's cubic may be defined by the parametric form

$$(1.22) \qquad x(t) = \sqrt{3}\,(t^2 - 1), \qquad y(t) = t\,(t^2 - 1),$$

corresponding to a special location of the origin, orientation, and scaling of the coordinate axes, and a specific parameterization (see [18]). To see that the cubic (1.15) we constructed by taking the rational parameterization (1.12) of the circle as our Pythagorean hodograph is equivalent to (1.22), we map (1.15) into (1.22) by: (i) taking $(x_0, y_0) = (0, -3)$, (ii) imposing the parameter transformation $t \to -\sqrt{3}\,t$, and (iii) interchanging $x(t)$ with $y(t)$ and shrinking both by a factor of $\sqrt{3}$.

The identification of Tschirnhausen's curve as the *unique* cubic with a Pythagorean hodograph in [18] is apparently new. Earlier authors were interested in this curve for rather different reasons; it can be shown, for example, that (1.15) is the *caustic curve* [39] for the reflection of parallel light rays by a parabola when those rays are incident normal to the axis of the parabola (the caustic curve is the "envelope" of the reflected rays[1](see Figs. 1.4 and 1.5).

---
[1]Salmon [35] attributes the first study of caustics to Tschirnhausen in 1682. An interesting discussion of these curves may be found in [3].

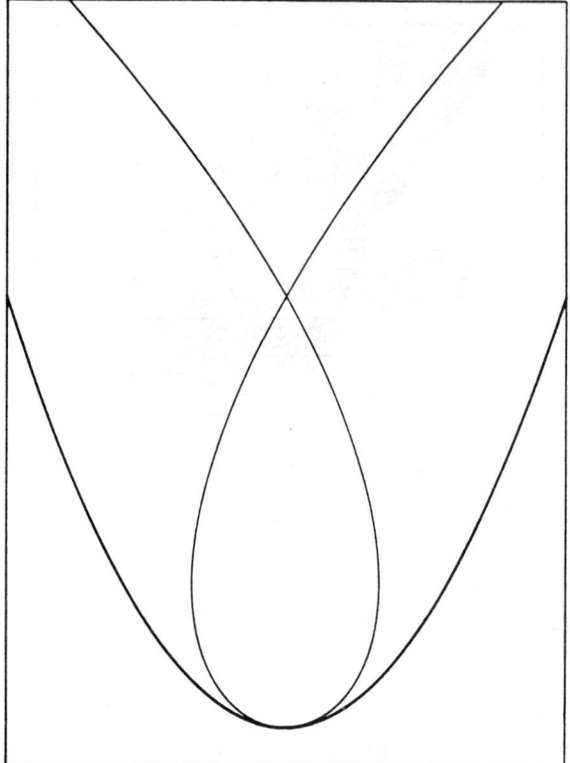

FIG. 1.5. *Tschirnhausen's cubic as a caustic curve.*

All Pythagorean–hodograph cubics with control points of the form (1.19) can be reduced to segments of the curve (1.22) under suitable displacements, rotations, uniform scalings, and linear reparameterizations. It is clear from Figs. 1.2 and 1.5 that Pythagorean–hodograph cubics are necessarily *convex*, and thus relinquish an important feature of general cubics: the ability to interpolate arbitrary sequences of points in the plane by C^2 piecewise cubic curves, i.e., cubic splines.

Since there is essentially just *one* Pythagorean–hodograph cubic, it is to be hoped that the quintics provide a greater degree of flexibility. To obtain a quintic Pythagorean–hodograph curve, we will choose the polynomials $u(t)$ and $v(t)$ in (1.16) to be quadratic:

(1.23)
$$u(t) = u_0(1-t)^2 + u_1 2(1-t)t + u_2 t^2,$$
$$v(t) = v_0(1-t)^2 + v_1 2(1-t)t + v_2 t^2,$$

and on integrating (1.16) we find that the control points of such curves are of the form

$$\mathbf{p_1} = \mathbf{p_0} + \frac{1}{5}(u_0^2 - v_0^2, 2u_0 v_0),$$

$$\mathbf{p_2} = \mathbf{p_1} + \frac{1}{5}(u_0 u_1 - v_0 v_1, u_0 v_1 + u_1 v_0),$$

$$(1.24) \; \mathbf{p}_3 \;\; = \;\; \mathbf{p}_2 + \frac{2}{15}(u_1^2 - v_1^2, 2u_1v_1) + \frac{1}{15}(u_0u_2 - v_0v_2, u_0v_2 + u_2v_0),$$

$$\mathbf{p}_4 \;\; = \;\; \mathbf{p}_3 + \frac{1}{5}(u_1u_2 - v_1v_2, u_1v_2 + u_2v_1),$$

$$\mathbf{p}_5 \;\; = \;\; \mathbf{p}_4 + \frac{1}{5}(u_2^2 - v_2^2, 2u_2v_2),$$

the point $\mathbf{p}_0$ being again arbitrary. The control points (1.24) define the lowest order Pythagorean–hodograph curves that may have real inflection points (see [18]) (another type of Pythagorean–hodograph quintic is obtained by choosing $u(t)$ and $v(t)$ as in (1.18) and a *quadratic* polynomial, rather than a constant, for $w(t)$; these curves are not as useful).

We are currently unaware of any simple answer to the matter of "how many" Pythagorean–hodograph quintics there are, and it seems probable that a lengthy analysis will be required to furnish that answer. Likewise, there do not appear to be any simple and intuitive geometric constraints on the control polygons of Bézier quintics — analogous to (1.21) in the case of cubics — which amount to sufficient and necessary conditions for a Pythagorean hodograph [18].

We will therefore address all concerns regarding the flexibility and construction of the Pythagorean–hodograph quintics by appealing to their most probable mode of use in practice, namely, as Hermite interpolants pieced together to form smooth "spline" curves.[2]

The Pythagorean–hodograph quintics are similar in certain respects to the regular parametric cubics: they may change their sense of curvature and they can satisfy the arbitrary first-order Hermite interpolation problem of matching specified end points and derivative values at those end points. These characteristics suggest that they might constitute a viable medium for practical design problems, rather than just a mathematical curiosity.

1.4. Hermite Interpolation

As is well known, there is a unique cubic arc that fits given end points $\mathbf{r}(0)$ and $\mathbf{r}(1)$, and derivatives $\mathbf{r}'(0)$ and $\mathbf{r}'(1)$ at those end points. Because of the constraints (1.21) on their control polygons, Pythagorean–hodograph cubic arcs clearly cannot (in general) satisfy such data. However, if we consider the *magnitudes* of the derivatives $\mathbf{r}'(0)$ and $\mathbf{r}'(1)$ immaterial, it may be possible to find segments of Tschirnhausen's cubic, suitably scaled and positioned, that interpolate prescribed end points for an arc and tangent *directions* at those points.

Using such interpolants, we might seek to fit sequences of points in the plane by tangent-continuous curves composed of Pythagorean–hodograph cubic arcs, invoking ad hoc procedures for assigning tangent directions at those

---

[2]We use the term spline rather loosely here; formally it connotes a function or parametric curve smoothly interpolating a sequence of points *and* minimizing a certain "strain energy" integral (see, for example, [7]).

points. Such curves are unlikely to satisfy engineering requirements, however, so we will not pursue them here.

Let us proceed instead to the problem of Hermite interpolation by Pythagorean–hodograph quintics. Since the first parametric derivatives at the ends of the Bézier curve (17) are simply

$$(1.25) \qquad \mathbf{r}'(0) = n\Delta\mathbf{p}_0 \quad \text{and} \quad \mathbf{r}'(1) = n\Delta\mathbf{p}_{n-1},$$

where $\Delta\mathbf{p}_0 = \mathbf{p}_1 - \mathbf{p}_0$ and $\Delta\mathbf{p}_{n-1} = \mathbf{p}_n - \mathbf{p}_{n-1}$, we see that the first-order Hermite interpolation problem of matching arbitrary end points and end derivatives corresponds to freely choosing the first and last two control points, $\mathbf{p}_0, \mathbf{p}_1$ and $\mathbf{p}_4, \mathbf{p}_5$.

Once these points have been specified to suit the Hermite interpolation problem under consideration, the coefficients (u_0, v_0) and (u_2, v_2) in (1.24) are determined in terms of them by the formulae [13]

$$(1.26) \qquad \begin{aligned} (u_0, v_0) &= \pm\sqrt{\frac{5}{2}}\left(\sqrt{|\Delta\mathbf{p}_0| + \Delta x_0}, \ \operatorname{sign}(\Delta y_0)\sqrt{|\Delta\mathbf{p}_0| - \Delta x_0}\right), \\ (u_2, v_2) &= \pm\sqrt{\frac{5}{2}}\left(\sqrt{|\Delta\mathbf{p}_4| + \Delta x_4}, \ \operatorname{sign}(\Delta y_4)\sqrt{|\Delta\mathbf{p}_4| - \Delta x_4}\right). \end{aligned}$$

Given these values of (u_0, v_0) and (u_2, v_2), we can determine the values of (u_1, v_1) appropriate to a Pythagorean–hodograph quintic in terms of the quantities

$$(1.27) \qquad \begin{aligned} A &= \frac{15}{2}(x_4 - x_1) + \frac{9}{16}(u_0^2 + u_2^2) - \frac{9}{16}(v_0^2 + v_2^2) + \frac{5}{8}(u_0 u_2 - v_0 v_2), \\ B &= \frac{15}{2}(y_4 - y_1) + \frac{9}{8}(u_0 v_0 + u_2 v_2) + \frac{5}{8}(u_0 v_2 + u_2 v_0), \end{aligned}$$

and $C = \sqrt{A^2 + B^2}$, by the formula

$$(1.28) \ (u_1, v_1) = -\frac{3}{4}(u_0 + u_2, v_0 + v_2) \pm \frac{1}{\sqrt{2}}\left(\sqrt{C + A}, \ \operatorname{sign}(B)\sqrt{C - A}\right).$$

Note that in (1.26) and (1.28) we adopt the convention

$$(1.29) \qquad \operatorname{sign}(z) = \begin{cases} -1 & \text{if } z < 0, \\ \pm 1 & \text{if } z = 0, \\ +1 & \text{if } z > 0. \end{cases}$$

A number of examples are illustrated in Figs. 1.6a–c; in each case we show the end points and derivative vectors to be interpolated, the Pythagorean–hodograph quintic interpolant (the bold curve) and its control polygon and, for purposes of comparison, the corresponding cubic Hermite interpolant (the light curve). The derivatives have been shortened by a factor of 5 in these illustrations.

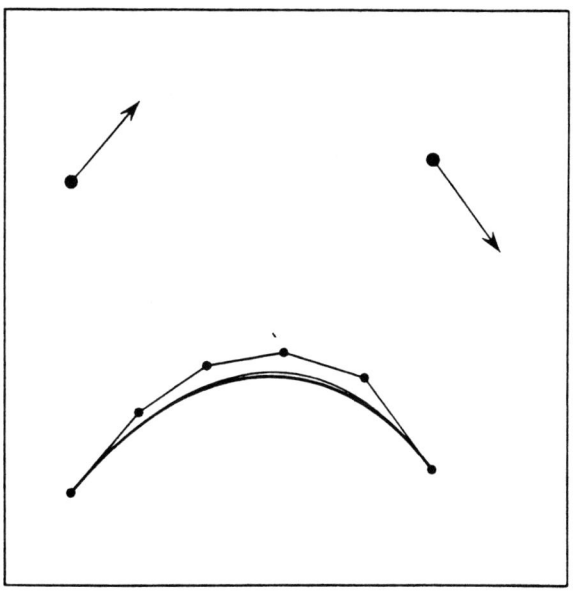

FIG. 1.6a. *A Pythagorean–hodograph quintic Hermite interpolant.*

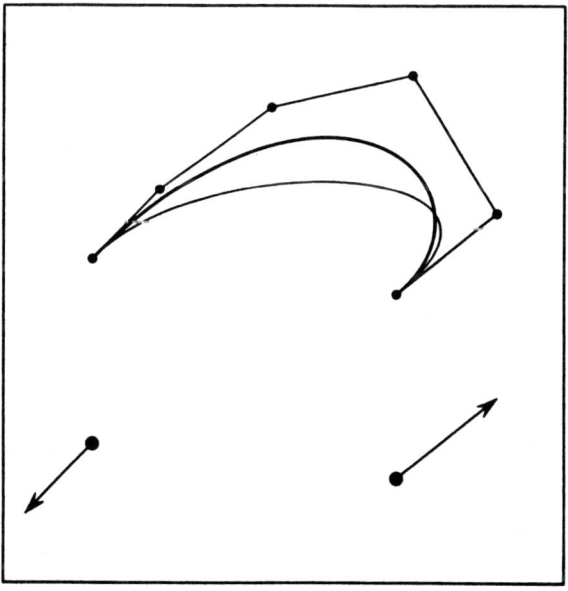

FIG. 1.6b. *Another Pythagorean–hodograph quintic Hermite interpolant.*

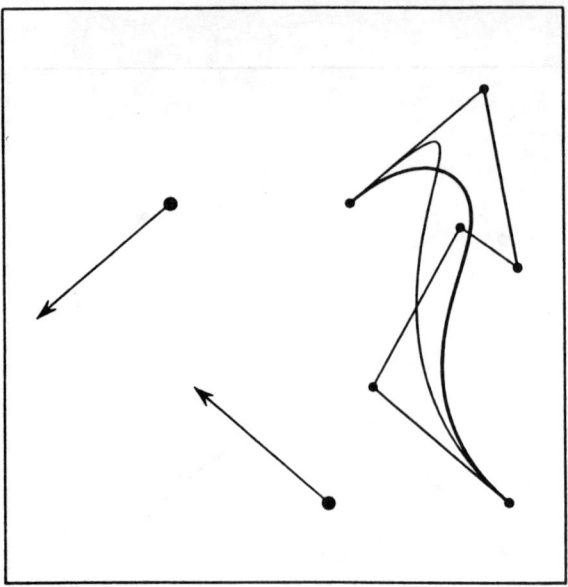

FIG. 1.6c. *A third Pythagorean–hodograph quintic Hermite interpolant.*

We have found that when the prescribed Hermite data is consistent with a smooth arc without inflections or strong curvature variations, the regular cubic and Pythagorean–hodograph quintic interpolants are very similar. Otherwise, the Pythagorean–hodograph quintic produces a somewhat "fairer" arc than the regular cubic, having a more even distribution of curvature. These traits are evident in the examples of Figure 1.6.

Now the astute reader will have noticed that, according to the freedom in choosing the signs in expressions (1.26) and (1.28) for the coefficients of the polynomials (1.23), there should actually be *more than one* interpolant to a given choice of $\mathbf{p}_0, \mathbf{p}_1$ and $\mathbf{p}_4, \mathbf{p}_5$. By substituting (1.26) and (1.28) into expressions (1.24) for the control points, it is not difficult to verify that there are in fact *four* distinct interpolants to each such choice.

However, our experiments suggest that one of these interpolants is usually much more "reasonable" than the others, since it does not exhibit strong curvature variations and/or loops at either or both ends of the arc (see Fig. 1.7) (the "reasonable" case was also chosen for each of the examples of Fig. 1.6).

Now from the usual expression for the curvature of a plane curve

$$(1.30) \qquad \kappa(t) = \frac{[\mathbf{r}'(t) \times \mathbf{r}''(t)] \cdot \mathbf{z}}{|\mathbf{r}'(t)|^3}$$

(where $\mathbf{z}$ is a unit vector orthogonal to the plane of $\mathbf{r}(t)$), we find that the end-point curvatures of the Pythagorean–hodograph quintic defined by (1.24) are given by

$$(1.31) \qquad \kappa(0) = 4\,\frac{u_0 v_1 - u_1 v_0}{(u_0^2 + v_0^2)^2} \quad \text{and} \quad \kappa(1) = 4\,\frac{u_1 v_2 - u_2 v_1}{(u_2^2 + v_2^2)^2}\,.$$

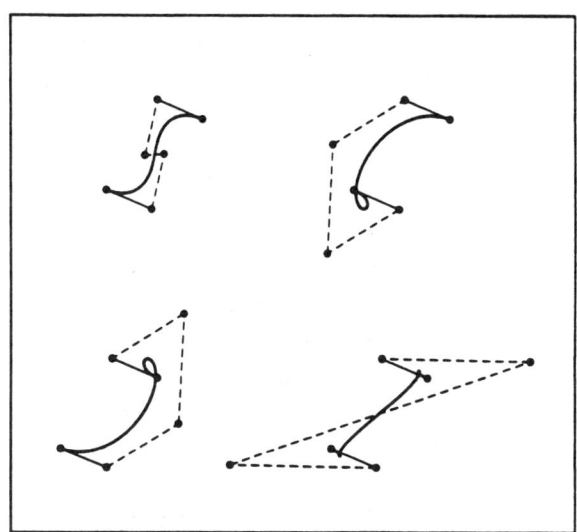

FIG. 1.7. *The four possible Pythagorean–hodograph quintics consistent with fixed control points* p_0, p_1 *and* p_4, p_5.

We assume that the coefficients on the right-hand side of (1.31) are given by any choice among the expressions (1.26)–(1.28) corresponding to a Hermite interpolant whose initial and final two control points are prescribed.

For comparison, the end-point curvatures of a cubic with control points p_0, p_1, p_2, p_3 are

$$(1.32) \qquad \kappa(0) = \frac{2}{3} \frac{\Delta p_0 \times \Delta p_1}{|\Delta p_0|^3} \quad \text{and} \quad \kappa(1) = \frac{2}{3} \frac{\Delta p_1 \times \Delta p_2}{|\Delta p_2|^3},$$

and if these control points have been chosen to satisfy the same Hermite interpolation problem such as the Pythagorean–hodograph quintic, we might hope that the values (1.31) and (1.32) would be in reasonable agreement.

In fact, exercising the sign freedoms in (1.26) and (1.28) alters, in general, both the magnitudes and signs of the end-point curvatures (1.31). When (as occurred often in practice and is the case in Fig. 1.7) only *one* of the four possible Pythagorean–hodograph quintics agrees with the corresponding cubic Hermite interpolant in the *signs* of its end-point curvatures, that one may clearly be uniquely identified as the most "reasonable."

If more than one — or even none! — of the four choices agree with the sense of the cubic end-point curvatures, a more detailed analysis is required, taking account of the magnitudes as well as the signs of the end-point curvatures (such cases typically arise when those curvatures are very weak). We defer a comprehensive examination of this problem to [13]; for the present we simply warn the reader that manual intervention may occasionally be required to adjust the choice of signs in (1.26) and (1.28) by trial and error to the most satisfactory combination.

Finally, we note another interesting possibility: using Pythagorean–hodograph quintics as second-order "geometric" Hermite interpolants (see [8]), matching prescribed end points, tangent *directions*, and curvatures. We have not explored this problem in any depth.

1.5. Pythagorean–Hodograph "Splines"

Let us turn now to the matter of smoothly interpolating a *sequence* of points in the plane by piecewise Pythagorean–hodograph curves. We begin by recalling some elementary facts concerning C^2 cubic splines.

Suppose that we wish to interpolate a sequence $r_0, \cdots, r_N$ of $N+1$ points in the plane by a C^2 piecewise cubic curve. Each cubic span k of this curve must interpolate consecutive points r_{k-1} and r_k and have first- and second-order parametric derivatives in agreement with those of its neighboring spans $k-1$ and $k+1$ at the points r_{k-1} and r_k.

The usual approach is to consider the constituent cubic arcs of the spline in Hermite form, regarding the first parametric derivatives $v_0, \cdots, v_N$ at the points $r_0, \cdots, r_N$ as unknowns that are determined by the second-order continuity requirement (see [7] or [9]).

If we assume a uniform knot vector for simplicity (i.e., that the points $r_0, r_1, \cdots, r_N$ are assigned parameter values $t = 0, 1, \cdots, N$), then matching the second derivative at the end of span k with that at the beginning of span $k+1$ gives rise to a tridiagonal system of (vector-valued) linear equations for the first derivatives $\{v_k\}$:

$$(1.33) \qquad v_{k-1} + 4\,v_k + v_{k+1} = 3\left(r_{k+1} - r_{k-1}\right)$$

for $k = 1, \cdots, N-1$. Since there are two more unknowns than equations, we need to impose additional "end conditions" in order to solve the system (1.33). We will not discuss the many possible choices here (see [9]), but assume simply that v_0 and v_N are to be chosen ad hoc.

Once the derivative vectors $\{v_k\}$ have been solved for, we can express each span k of the spline curve in Bernstein–Bézier form with control points

$$(1.34) \quad p_0 = r_{k-1}, \quad p_1 = r_{k-1} + \frac{1}{3}v_{k-1}, \quad p_2 = r_k - \frac{1}{3}v_k, \quad p_3 = r_k$$

for $k = 1, \cdots, N$.

One possible approach to constructing C^1 "splines" composed of Pythagorean–hodograph quintic arcs is to simply solve the system (1.33) for the derivative values v_k — as though we were constructing a C^2 cubic spline — and then construct the Pythagorean–hodograph quintic interpolant to the Hermite data (r_{k-1}, v_{k-1}) and (r_k, v_k) on each span k. Specifically, we assign the first and last two control points by

$$(1.35) \quad p_0 = r_{k-1}, \quad p_1 = r_{k-1} + \frac{1}{5}v_{k-1}, \quad p_4 = r_k - \frac{1}{5}v_k, \quad p_5 = r_k,$$

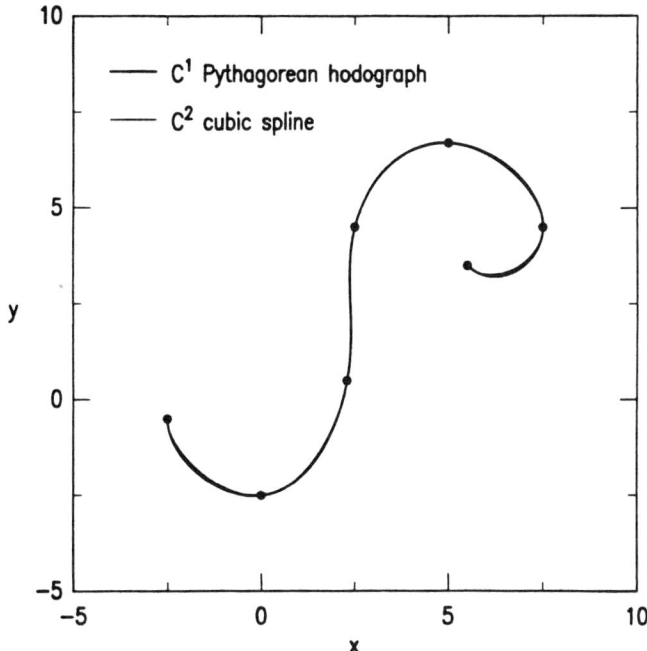

FIG. 1.8. *Comparison of a C^1 Pythagorean–hodograph "spline" and the traditional C^2 cubic spline interpolating seven points.*

and then compute the appropriate locations of the interior control points $\mathbf{p}_2$ and $\mathbf{p}_3$ from equations (1.26)–(1.28) and (1.24).

Notwithstanding the fact that the above scheme relinquishes precise curvature continuity[3] it was found in practice to produce curves very similar to the corresponding C^2 cubic splines.

Figure 1.8 illustrates an example interpolating seven points; the bold curve is the C^1 Pythagorean–hodograph curve and the light curve is the traditional C^2 cubic spline. Despite the convoluted shape of the underlying data, the two curves are virtually indistinguishable! The difference is somewhat more evident if we compare curvatures (Fig. 1.9), although the discontinuities that the Pythagorean–hodograph curve suffers are relatively mild.

The C^1 Pythagorean–hodograph interpolation scheme described above is similar in spirit to various "shape-preserving" methods (e.g., [22], [23]) that rely on interpolating tangent data selected a priori such as to ensure that the interpolant inherits certain desirable characteristics (e.g., monotonicity or convexity) of the point data.

If we *insist* on C^2 continuity, the piecewise Pythagorean–hodograph quintic interpolation problem is not trivial, requiring the solution of a fairly large system of coupled quadratic equations. We will adopt an approach similar

---

[3]Although curvature continuity is essential for certain applications, the attention that it (or even higher orders of continuity) receives in the literature is, in the author's opinion, motivated primarily by academic rather than practical considerations. It takes a trained eye to even *see* quite substantial jumps in curvature along a merely tangent-continuous curve.

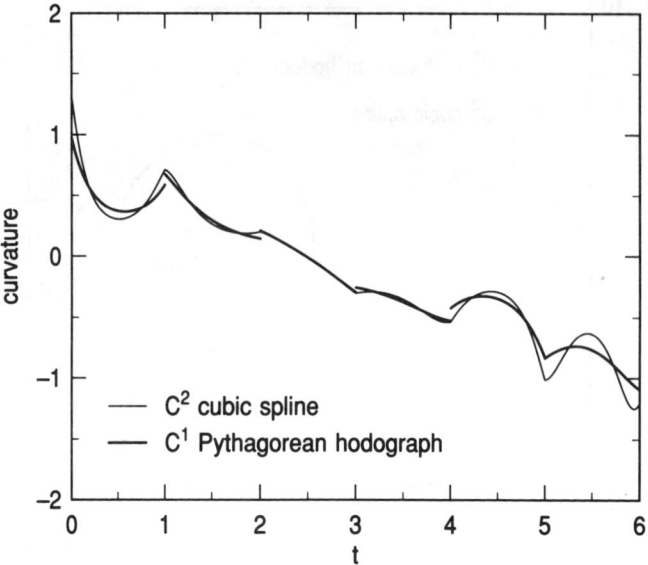

FIG. 1.9. *Curvature variation along the splines of Fig. 1.8.*

to that used for regular cubic splines. At each point $\mathbf{r}_k$ we assign a pair of indeterminates (λ_k, μ_k) for $k = 0, \cdots, N$. Similarly, with each span k between consecutive points $\mathbf{r}_{k-1}$ and $\mathbf{r}_k$ we associate another pair of indeterminates (α_k, β_k) for $k = 1, \cdots, N$.

These indeterminates are to be interpreted as follows. On span k, the Pythagorean–hodograph quintic arc is determined by polynomials $u(t)$ and $v(t)$ of the form (1.23), with coefficients given by

$$(1.36) \quad (u_0, v_0) = (\lambda_{k-1}, \mu_{k-1}), \quad (u_1, v_1) = (\alpha_k, \beta_k), \quad (u_2, v_2) = (\lambda_k, \mu_k).$$

Since from (1.24) and (1.25) the end-point derivatives of the kth arc are

$$(1.37) \quad \mathbf{r}'(0) = (u_{k-1}^2 - v_{k-1}^2, 2u_{k-1}v_{k-1}) \quad \text{and} \quad \mathbf{r}'(1) = (u_k^2 - v_k^2, 2u_k v_k),$$

it is clear that consecutive arcs are necessarily matched to first order.

By consideration of second end-point derivatives, we then find that the equations

$$(1.38a) \qquad 2(\lambda_k^2 - \mu_k^2) - (\alpha_k + \alpha_{k+1})\lambda_k + (\beta_k + \beta_{k+1})\mu_k = 0,$$

$$(1.38b) \qquad 4\lambda_k \mu_k - (\beta_k + \beta_{k+1})\lambda_k - (\alpha_k + \alpha_{k+1})\mu_k = 0,$$

for $k = 1, \cdots, N-1$, correspond to the imposition of second-order continuity at each of the $N-1$ "interior" points $\mathbf{r}_1, \cdots, \mathbf{r}_{N-1}$.

In order to ensure that the quintic arcs are indeed of Pythagorean–hodograph form, the following equations must also be satisfied:
(1.39a)
$$3(\lambda_{k-1}^2 - \mu_{k-1}^2) + 3\lambda_k^2 - \mu_k^2) + 3(\alpha_k \lambda_k - \beta_k \mu_k) + 3(\alpha_k \lambda_{k-1} - \beta_k \mu_{k-1})$$
$$+ 2(\alpha_k^2 - \beta_k^2) + (\lambda_{k-1}\lambda_k - \mu_{k-1}\mu_k) = 15(x_k - x_{k-1}),$$

(1.39b)
$$6\lambda_{k-1}\mu_{k-1} + 6\lambda_k\mu_k + 3(\alpha_k\mu_k + \beta_k\lambda_k) + 3(\alpha_k\mu_{k-1} + \beta_k\lambda_{k-1})$$
$$+ 4\alpha_k\beta_k + (\lambda_{k-1}\mu_k + \lambda_k\mu_{k-1}) = 15(y_k - y_{k-1}),$$

on each span $k = 1, \cdots, N$.

Altogether, (1.38a,b) and (1.39a,b) constitute a system of $4N - 2$ coupled quadratic equations for the $4N + 2$ unknowns $\{\lambda_k, \mu_k\}$ for $k = 0, \cdots, N$ and $\{\alpha_k, \beta_k\}$ for $k = 1, \cdots, N$. Thus, as in the regular cubic interpolating spline case, we need additional end conditions for this system. We might ask, for example, that the quantities (λ_0, μ_0) and (λ_N, μ_N) be arbitrarily chosen, corresponding to the imposition of prescribed parametric derivative values at the first and last point.

(Note that in the regular cubic spline case the x and y components of the vector tridiagonal system (1.33) are solved independently, whereas in the Pythagorean–hodograph case such a decoupling does not exist. Thus, the nonlinear nature of (1.38a,b) and (1.39a,b) defining a C^2 Pythagorean–hodograph spline is compounded with the fact that we are dealing with a *much larger* system of equations than in the regular cubic spline case.)

We have not attempted to implement this C^2 interpolation scheme nor do we claim to be certain that (1.38a,b) and (1.39a,b) can be relied upon to yield real solutions for *arbitrary* sequences of points $\mathbf{r}_0, \cdots, \mathbf{r}_N$ and choices of (λ_0, μ_0) and (λ_N, μ_N). In view of our preceding footnote concerning the practical significance of precise curvature continuity, and the remarkable similarity of regular C^2 cubic splines and their Pythagorean–hodograph quintic C^1 counterparts, such an exercise would seem to be of primarily academic interest.

1.6. Parametric Speed

Recall that the parametric speed of a Pythagorean–hodograph curve $\mathbf{r}(t) = \{x(t), y(t)\}$ of the form (1.16) is given by the polynomial

$$(1.40) \qquad \sigma(t) = \sqrt{x'^2(t) + y'^2(t)} = u^2(t) + v^2(t).$$

If $\mathbf{r}(t)$ is of (odd) degree n, then $u(t)$ and $v(t)$ must be of degree

$$(1.41) \qquad m = \tfrac{1}{2}(n - 1),$$

and may be written in the form

$$(1.42) \quad u(t) = \sum_{k=0}^{m} u_k \binom{m}{k} (1-t)^{m-k} t^k, \qquad v(t) = \sum_{k=0}^{m} v_k \binom{m}{k} (1-t)^{m-k} t^k.$$

We wish to express (1.40) in the Bernstein–Bézier form

$$(1.43) \qquad \sigma(t) = \sum_{k=0}^{n-1} \sigma_k \binom{n-1}{k} (1-t)^{n-1-k} t^k,$$

in terms of the cofficients $\{u_k\}$ and $\{v_k\}$ of the polynomials (1.42). By means of the multiplication procedure for polynomials in Bernstein form [17], we have

$$(1.44) \qquad \sigma_k = \sum_{j=\max(0,k-m)}^{\min(m,k)} \frac{\binom{m}{j}\binom{m}{k-j}}{\binom{n-1}{k}} (u_j u_{k-j} + v_j v_{k-j})$$

for $k = 0, \cdots, n-1$.

Thus, for the Pythagorean–hodograph cubics, the coefficients of $\sigma(t)$ are given by

$$(1.45) \qquad \begin{aligned} \sigma_0 &= u_0^2 + v_0^2, \\ \sigma_1 &= u_0 u_1 + v_0 v_1, \\ \sigma_2 &= u_1^2 + v_1^2, \end{aligned}$$

while for the quintics we have

$$(1.46) \qquad \begin{aligned} \sigma_0 &= u_0^2 + v_0^2, \\ \sigma_1 &= u_0 u_1 + v_0 v_1, \\ \sigma_2 &= \frac{2}{3}(u_1^2 + v_1^2) + \frac{1}{3}(u_0 u_2 + v_0 v_2), \\ \sigma_3 &= u_1 u_2 + v_1 v_2, \\ \sigma_4 &= u_2^2 + v_2^2. \end{aligned}$$

In order to integrate the parametric speed (1.43) and thus obtain the arc length as a polynomial function of the parameter t,

$$(1.47) \qquad s(t) = \int_0^t \sigma(\tau)\,d\tau,$$

we make use of the following rule [17]:

$$(1.48) \qquad \int \binom{n-1}{k}(1-t)^{n-1-k}t^k\,dt = \frac{1}{n}\sum_{j=k+1}^n \binom{n}{j}(1-t)^{n-j}t^j,$$

$k = 0, \cdots, n-1$, for the indefinite integrals of the Bernstein basis functions. This gives

$$(1.49) \qquad s(t) = \sum_{k=0}^n s_k \binom{n}{k}(1-t)^{n-k}t^k,$$

where $s_0 = 0$ and

$$(1.50) \qquad s_k = \frac{1}{n}\sum_{j=0}^{k-1} \sigma_j \quad \text{for } k = 1, \cdots, n.$$

Hence the total arc length S is just

(1.51) $$S = s(1) = \frac{\sigma_0 + \sigma_1 + \cdots + \sigma_{n-1}}{n}.$$

Thus, in order to compute the arc length of any parametric span $[0, t]$ of a Pythagorean–hodograph curve, we need only evaluate the polynomial (1.49) at t. The simplicity of this computation may be compared with that of obtaining the value of the integral

(1.52) $$s(t) = \int_0^t \sqrt{x'^2(\tau) + y'^2(\tau)}\, d\tau$$

by means of numerical quadrature, since for general polynomial curves this integral cannot be resolved into elementary functions of t (see, for example, [24]).

Likewise, it is much simpler to determine from (1.49) than from (1.52) the parameter value t_0 at which the arc length, measured from $t = 0$, attains a prescribed value s_0, i.e., to solve the equation

(1.53) $$s(t_0) = s_0,$$

with $s(t)$ given by (1.49) or (1.52). By way of illustration, let us consider the problem of *uniformly rendering* a parametric curve.

It is common practice in CAGD and graphics applications to render a parametric curve $\mathbf{r}(t)$ by evaluating at a sequence $t_0, \cdots, t_N$ of values corresponding to a uniform increment $\Delta t = t_k - t_{k-1}$ for $k = 1, \cdots, N$ in the parameter. However, this leads to an uneven spacing (by arc length) of the geometric points $\mathbf{r}(t_k)$ on the curve, since its parametric speed $\sigma(t)$ is not, in general, constant.[4]

If we desire a uniformly spaced sequence of $N + 1$ points on a general polynomial curve $\mathbf{r}(t)$, corresponding to a constant increment $\Delta s = S/N$ in the arc length between consecutive points, we must determine for $k = 1, \cdots, N - 1$ when the integral (1.52), *considered as a function of the upper limit of integration*, attains the desired values $k\,\Delta s$. (The set of points defined by $t_k = t_{k-1} + \Delta s / |\mathbf{r}'(t_{k-1})|$ for $k = 1, \cdots, N - 1$ with $t_0 = 0$ and $t_N = 1$ offer a crude approximation, but refining them is a numerically intensive and somewhat awkward problem.)

Although the parametric speed of a Pythagorean–hodograph curve is also nonconstant, the particularly simple form (1.49) of $s(t)$ makes it a rather easy matter to compensate for its variation. Let the parameter values of the points uniformly spaced by arc length be $\{t_k\}$, so that

(1.54) $$s(t_k) = k\,\Delta s$$

---

[4]In fact, this nonuniformity is an intrinsic defect of all "simple" (polynomial or rational) curve parameterizations. It may be shown [19] that the parametric speed of *any* such curve — other than a straight line — is necessarily nonconstant.

for $k = 1, \cdots, N-1$, with $t_0 = 0$ and $t_N = 1$. Note that, since $\sigma(t) = ds/dt$ and $\sigma(t) > 0$ for all real t when $\mathrm{GCD}(u, v) = 1$, $s(t)$ is strictly monotone increasing with t, and for each k the polynomial (1.54) has only one simple real root.

(The monotonicity of $s(t)$ implies the existence of an inverse function $t(s)$ giving the parameter in terms of the arc length [4]. By the preceding footnote, $t(s)$ is obviously not a simple (rational) function, but the form (1.49) offers a useful point of departure for approximating it over a finite range of t, and thereby obtaining approximations to the *intrinsic parameterization* of $\mathbf{r}(t)$ through the substitution $\mathbf{r}(s) = \mathbf{r}(t(s))$. See also [38].)

Clearly, the real root t_k of the equation $s(t) - k\,\Delta s = 0$ must lie between t_{k-1} and 1. As an initial approximation to it, we take

$$(1.55) \qquad\qquad \tau_0 = t_{k-1} + \frac{\Delta s}{\sigma(t_{k-1})}$$

and obtain further refinements by applying the Newton–Raphson iteration

$$(1.56) \qquad \tau_{r+1} = \tau_r - \frac{s(\tau_r) - k\Delta s}{\sigma(\tau_r)} \quad \text{for } r = 0, 1, \cdots.$$

As is well known [6], this iteration is *quadratically convergent* for starting approximations τ_0 sufficiently close to t_k, and in practical examples (see Fig. 1.10) the parameter values $\{t_k\}$ were obtained to an accuracy of 10^{-12} or better in just two or three iterations. Since *precise* uniformity by arc length is probably not crucial in most applications, a single iteration on the starting values (1.55) will often suffice. Typically, this gives uniform spacing to a relative accuracy of about 10^{-6}.

The reader may justifiably be concerned about whether the starting value (1.55) is always "sufficiently close" to guarantee convergence of the iteration (1.56) to the desired root. Traditional criteria for guaranteeing convergence involve verifying that the first and second derivatives do not change sign over an isolating interval for the root [25]. While the Bernstein–Bézier form offers a simple a priori means for this [12], the additional expense of such checks was found to be unwarranted in all the practical examples studied.

1.7. Offset Curves

We will now show how the offsets to Pythagorean–hodograph cubics and quintics can be formulated precisely as *rational* Bézier curves of degree 5 and 9, respectively. Let the control points of the Pythagorean–hodograph curve $\mathbf{r}(t)$ be written in homogeneous form as

$$(1.57) \qquad\qquad \mathbf{P}_k = (x_k, y_k, 1)$$

for $k = 0, \cdots, n$. The first forward differences of (1.57) are given by

$$(1.58) \qquad\qquad \Delta\mathbf{P}_k = \mathbf{P}_{k+1} - \mathbf{P}_k = (\Delta x_k, \Delta y_k, 0)$$

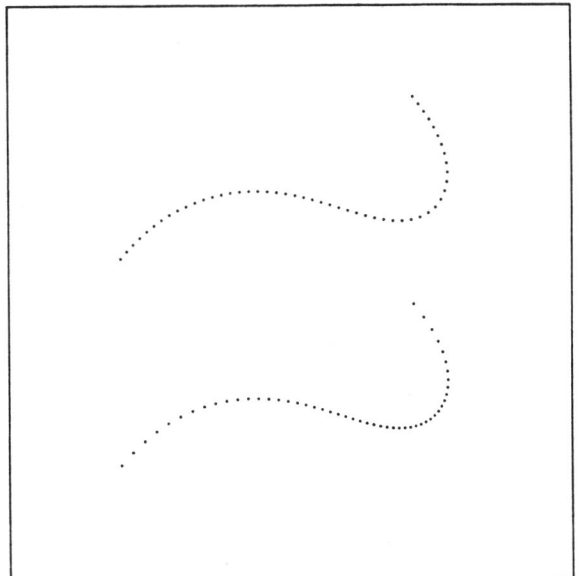

FIG. 1.10. *A Pythagorean–hodograph quintic curve, rendered by uniform arc length and parametric increments.*

for $k = 0, \cdots, n - 1$ (where $\Delta x_k = x_{k+1} - x_k$ and $\Delta y_k = y_{k+1} - y_k$), and introducing a unit vector $\mathbf{z}$ orthogonal to the plane $\mathbf{r}(t)$, we define the quantities

$$(1.59) \qquad \Delta \mathbf{P}_k \times \mathbf{z} = (\Delta y_k, -\Delta x_k, 0).$$

The offset to $\mathbf{r}(t)$ at distance d can be expressed as

$$(1.60) \qquad \mathbf{r}_o(t) = \mathbf{r}(t) + d\, \frac{\mathbf{r}'(t) \times \mathbf{z}}{|\mathbf{r}'(t)|}.$$

Thus, if we write $\mathbf{r}'(t)$ in the form

$$(1.61) \qquad \mathbf{r}'(t) = n \sum_{k=0}^{n-1} \Delta \mathbf{p}_k \binom{n-1}{k} (1-t)^{n-1-k} t^k,$$

and note that $|\mathbf{r}'(t)|$ is given by the *polynomial* $\sigma(t)$ in (1.43), we see that the offset (1.60) has the rational form

$$(1.62) \qquad \mathbf{r}_o(t) = \left(\frac{X(t)}{W(t)}, \frac{Y(t)}{W(t)} \right).$$

Here $X(t), Y(t), W(t)$ are polynomials of degree $2n - 1$, whose coefficients

$$(1.63) \qquad \mathbf{O}_k = (X_k, Y_k, W_k) \quad \text{for } k = 0, \cdots, 2n - 1$$

correspond to the Bézier control points of the rational offset curve.

The homogeneous coordinates (1.63) for the control points of the offset may be concisely expressed in terms of those (1.57) of the original curve as

$$(1.64) \quad \mathbf{O}_k = \sum_{j=\max(0,k-n)}^{\min(n-1,k)} \frac{\binom{n-1}{j}\binom{n}{k-j}}{\binom{2n-1}{k}} [\sigma_j \mathbf{P}_{k-j} + d\,n\Delta\mathbf{P}_j \times \mathbf{z}]$$

[18] for $k = 0, \cdots, 2n - 1$.

For Pythagorean–hodograph cubics, the control points of the rational quintic offset curves are thus given by

$$\mathbf{O}_0 = \sigma_0\mathbf{P}_0 + 3d\,\Delta\mathbf{P}_0 \times \mathbf{z},$$

$$\mathbf{O}_1 = \frac{1}{5}[2\sigma_1\mathbf{P}_0 + 3\sigma_0\mathbf{P}_1 + 3d\,(3\Delta\mathbf{P}_0 + 2\Delta\mathbf{P}_1)\times\mathbf{z}],$$

$$(1.65) \quad \mathbf{O}_2 = \frac{1}{10}[\sigma_2\mathbf{P}_0 + 6\sigma_1\mathbf{P}_1 + 3\sigma_0\mathbf{P}_2 + 3d\,(3\Delta\mathbf{P}_0 + 6\Delta\mathbf{P}_1 + \Delta\mathbf{P}_2)\times\mathbf{z}],$$

$$\mathbf{O}_3 = \frac{1}{10}[3\sigma_2\mathbf{P}_1 + 6\sigma_1\mathbf{P}_2 + \sigma_0\mathbf{P}_3 + 3d\,(\Delta\mathbf{P}_0 + 6\Delta\mathbf{P}_1 + 3\Delta\mathbf{P}_2)\times\mathbf{z}],$$

$$\mathbf{O}_4 = \frac{1}{5}[3\sigma_2\mathbf{P}_2 + 2\sigma_1\mathbf{P}_3 + 3d\,(2\Delta\mathbf{P}_1 + 3\Delta\mathbf{P}_2)\times\mathbf{z}],$$

$$\mathbf{O}_5 = \sigma_2\mathbf{P}_3 + 3d\,\Delta\mathbf{P}_2 \times \mathbf{z}.$$

For Pythagorean–hodograph quintics, the offsets are rational curves of degree 9 with control points

$$\mathbf{O}_0 = \sigma_0\mathbf{P}_0 + 5d\,\Delta\mathbf{P}_0 \times \mathbf{z},$$

$$\mathbf{O}_1 = \frac{1}{9}[4\sigma_1\mathbf{P}_0 + 5\sigma_0\mathbf{P}_1 + 5d\,(5\Delta\mathbf{P}_0 + 4\Delta\mathbf{P}_1)\times\mathbf{z}],$$

$$\mathbf{O}_2 = \frac{1}{18}[3\sigma_2\mathbf{P}_0 + 10\sigma_1\mathbf{P}_1 + 5\sigma_0\mathbf{P}_2 + 5d\,(5\Delta\mathbf{P}_0 + 10\Delta\mathbf{P}_1 + 3\Delta\mathbf{P}_2)\times\mathbf{z}],$$

$$\mathbf{O}_3 = \frac{1}{42}[2\sigma_3\mathbf{P}_0 + 15\sigma_2\mathbf{P}_1 + 20\sigma_1\mathbf{P}_2 + 5\sigma_0\mathbf{P}_3$$
$$+ 5d\,(5\Delta\mathbf{P}_0 + 20\Delta\mathbf{P}_1 + 15\Delta\mathbf{P}_2 + 2\Delta\mathbf{P}_3)\times\mathbf{z}],$$

$$\mathbf{O}_4 = \frac{1}{126}[\sigma_4\mathbf{P}_0 + 20\sigma_3\mathbf{P}_1 + 60\sigma_2\mathbf{P}_2 + 40\sigma_1\mathbf{P}_3 + 5\sigma_0\mathbf{P}_4$$

$$(1.66) \quad\quad\quad + 5d\,(5\Delta\mathbf{P}_0 + 40\Delta\mathbf{P}_1 + 60\Delta\mathbf{P}_2 + 20\Delta\mathbf{P}_3 + \Delta\mathbf{P}_4)\times\mathbf{z}],$$

$$\mathbf{O}_5 = \frac{1}{126}[5\sigma_4\mathbf{P}_1 + 40\sigma_3\mathbf{P}_2 + 60\sigma_2\mathbf{P}_3 + 20\sigma_1\mathbf{P}_4 + \sigma_0\mathbf{P}_5$$

$$+ 5d\,(\Delta\mathbf{P}_0 + 20\Delta\mathbf{P}_1 + 60\Delta\mathbf{P}_2 + 40\Delta\mathbf{P}_3 + 5\Delta\mathbf{P}_4)\times\mathbf{z}],$$

$$\mathbf{O}_6 = \frac{1}{42}[5\sigma_4\mathbf{P}_2 + 20\sigma_3\mathbf{P}_3 + 15\sigma_2\mathbf{P}_4 + 2\sigma_1\mathbf{P}_5$$

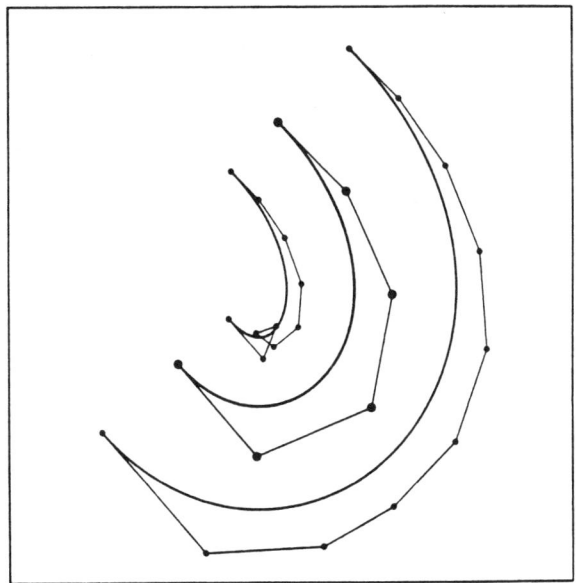

FIG. 1.11. *Control polygons for two degree 9 rational offsets to a Pythagorean–hodograph quintic.*

$$+5d\,(2\Delta\mathbf{P}_1 + 15\Delta\mathbf{P}_2 + 20\Delta\mathbf{P}_3 + 5\Delta\mathbf{P}_4)\times\mathbf{z}\,],$$

$$\mathbf{O}_7 = \frac{1}{18}\,[\,5\sigma_4\mathbf{P}_3 + 10\sigma_3\mathbf{P}_4 + 3\sigma_2\mathbf{P}_5 + 5d\,(3\Delta\mathbf{P}_2 + 10\Delta\mathbf{P}_3 + 5\Delta\mathbf{P}_4)\times\mathbf{z}\,],$$

$$\mathbf{O}_8 = \frac{1}{9}\,[\,5\sigma_4\mathbf{P}_4 + 4\sigma_3\mathbf{P}_5 + 5d\,(4\Delta\mathbf{P}_3 + 5\Delta\mathbf{P}_4)\times\mathbf{z}\,],$$

$$\mathbf{O}_9 = \sigma_4\mathbf{P}_5 + 5d\,\Delta\mathbf{P}_4\times\mathbf{z}\,.$$

Figure 1.11 illustrates the control polygons and corresponding offset curves defined by (1.66) for representative positive and negative values of d. Note that since the offsets are *rational* curves (each control point having in general a different projective coordinate or "weight" W_k), the shape of the control polygon is not necessarily a reliable indicator of the shape of the offset curve that it defines.

Expressions (1.66) — and (1.46) — are not difficult (only slightly tedious) to code. A more serious concern regarding them is the possibility that the compounding of inaccuracies from several sources in deriving and evaluating the final offset curve using floating point arithmetic might lead to a serious overall degradation in the reliability of this approach.

Among the sources of error are: (a) the rather involved expressions (1.24) and (1.26)–(1.28) giving the control points appropriate to prescribed Hermite data, the reliance on approximate root extractions resulting in only a "nearly Pythagorean" hodograph; (b) the considerable quantity of finite precision arithmetic required in (1.46) and (1.66) to obtain the offset control points; and (c) the nature of the offset itself, a rational form of degree 9, which might be expected to incur appreciable error in its evaluation.

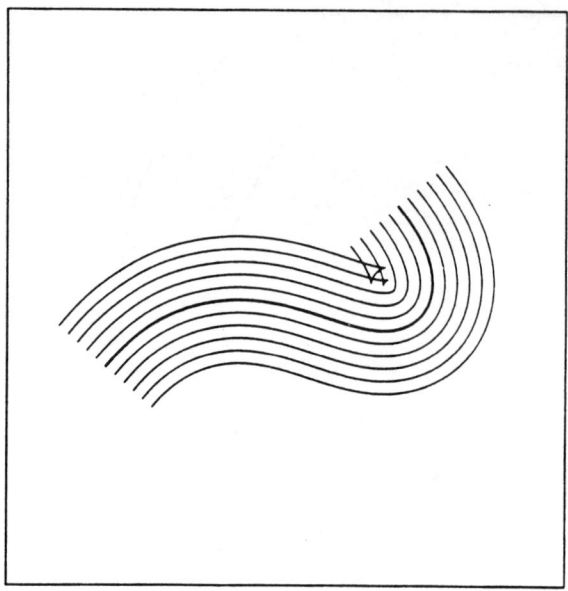

FIG. 1.12. *Exact rational offsets to a Pythagorean–hodograph quintic.*

In practice, these fears transpired to be quite unfounded. Figure 1.12 illustrates a family of offsets to one of the quintic Pythagorean–hodograph interpolants of Fig. 1.6. The interpolant and its rational degree 9 offsets were computed in standard double precision arithmetic. Sequences of corresponding points on the generator and offsets were then evaluated, and the measured distances between them were compared with the prescribed offset distance.

In all cases (even for the "degenerate" offsets with cusps and loops), the measured fractional error never exceeded 10^{-14}. Such accuracy — no doubt largely attributable to our strict adherence to the Bernstein–Bézier form — is virtually unattainable by approximation schemes when the amount of data is to be confined to reasonable bounds.

In Figs. 1.13a and 1.13b we show piecewise cubic approximations to the innermost and outermost offsets of Fig. 1.12, obtained by the (admittedly somewhat naïve) scheme of sampling the generator curve $\mathbf{r}(t)$ at $2^k + 1$ points $0, t_1, t_2, \cdots, 1$ uniformly spaced by parameter value, and then computing the cubic Hermite interpolants to the offset data

$$(1.67) \qquad \begin{aligned} \mathbf{r}_o(t_k) &= \mathbf{r}(t_k) + d\,\mathbf{n}(t_k), \\ \mathbf{r}'_o(t_k) &= [\,1 + \kappa(t_k)d\,]\,\mathbf{r}'(t_k), \end{aligned}$$

at adjacent points, for $k = 1, 2, \cdots$ (where $\mathbf{n}(t)$ and $\kappa(t)$ are the unit normal and curvature of $\mathbf{r}(t)$; see [14]).

Figure 1.14 illustrates the greatest relative errors arising in these offset approximations as the number of cubic segments (2^k) is increased, and the observed bound on the error of the Pythagorean–hodograph form (1.66) due to round-off (note that *each* cubic segment is defined by eight coefficients,

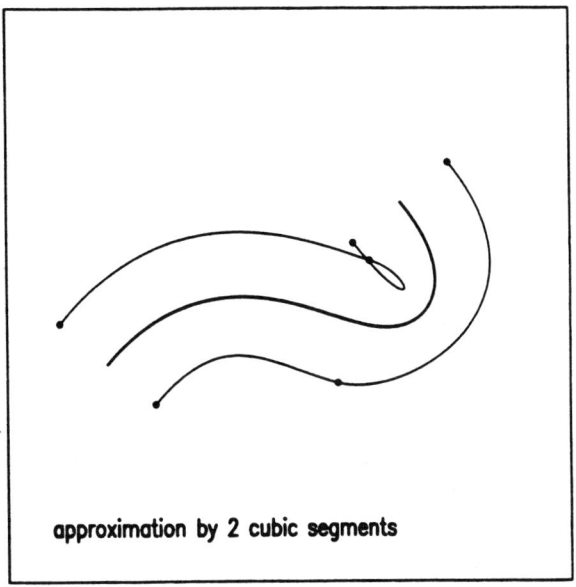

FIG. 1.13a. *Initial piecewise cubic offset approximation.*

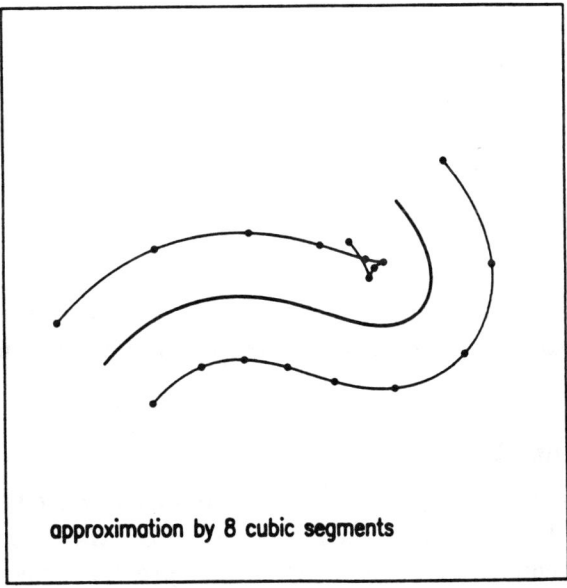

FIG. 1.13b. *Improved piecewise cubic offset approximation.*

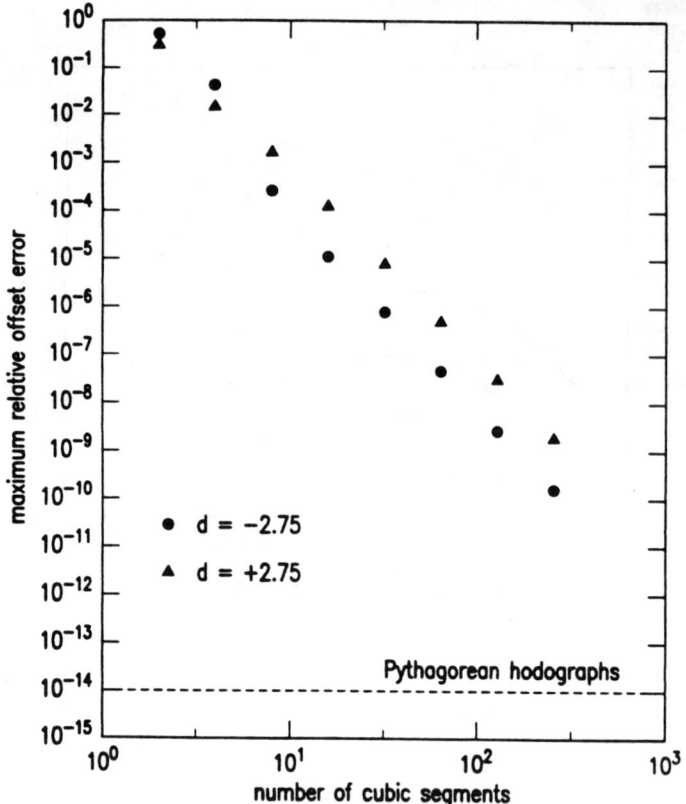

FIG. 1.14. *Maximum relative errors in the offset approximations.*

while the exact rational offset of degree 9 requires a total of 40). While the performance shown in Figure 1.14 can undoubtedly be improved by use of more intelligent subdivision strategies, its basic message is unequivocal: to achieve offset accuracies comparable to the Pythagorean–hodograph form, approximation schemes require vastly more data.

The rate of decrease of the offset approximation error evident in Figure 1.14 is consistent with the usual notion that the remainder term for cubic Hermite interpolation decreases as $O(h^4)$, where h is the sampling interval (see [7]); the constant factor of this rate is determined by the greatest magnitude of the fourth derivative of $\mathbf{r}_o(t)$ for $t \in [0, 1]$.

1.8. Concluding Remarks

We hope that the preceding discussion and illustrations have helped to convince the reader that Pythagorean–hodograph curves — the quintics, at least — offer a viable medium for engineering design problems and are not merely mathematical oddities. The proof of the pudding, of course, is in the eating, and we must confess that the examples we have tested thus far cannot be regarded as more than a few preliminary nibbles.

While the construction of Pythagorean–hodograph curves is undoubtedly more involved than that of regular parametric polynomial curves, the reader should bear in mind the great advantage of subsequently being able to compute, store, and manipulate offset curves *without approximation* while remaining within the established representational and algorithmic conventions of current Computer Aided Design (CAD) systems.[5] In specific applications that call for an extensive processing of two-dimensional offsets (e.g., numerical control machining) and do not require precise C^2 continuity, we believe that the Pythagorean–hodograph quintics offer a promising new approach.

How do the ideas presented herein extend to surfaces? Is there a class of relatively low degree parametric surfaces (beyond the cyclides [33]) whose offsets possess exact rational parameterizations, bypassing the difficulties and deficiencies of offset surface approximation schemes [11]? And if so, how do we go about constructing and using them?

Unfortunately, the extension to surfaces is by no means straightforward (see [18] for a precise statement of the problem). We will content ourselves here with the observation that surfaces defined by simple schemes for "blending" together Pythagorean–hodograph boundary curves (i.e., "tensor product" patches) do *not* have the desired property.

Finally let us mention that, although we have skimped on details, the formulae enumerated herein offer the reader an adequate point of departure for trying out the Pythagorean–hodograph curves against practical design problems, to see whether or not he or she agrees with our assessment. In either case, the author would appreciate hearing of such experiences.

References

[1] P. Bézier, *Numerical Control: Mathematics and Applications*, John Wiley, London, 1972.

[2] J. Bronowski, *The Ascent of Man*, British Broadcasting Corporation, London, 1973.

[3] J. W. Bruce, P. J. Giblin, and C. G. Gibson, *On caustics of plane curves*, Amer. Math. Monthly, 88 (1981), pp. 651–667.

[4] R. C. Buck, *Advanced Calculus* (3rd Edition), McGraw-Hill, New York, 1978.

[5] S. Coquillart, *Computing offsets of B-spline curves*, Comput. Aided Des., 19 (1987), pp. 35–39.

[6] G. Dahlquist and Å. Björck, *Numerical Methods*, Prentice-Hall, Englewood Cliffs, NJ, 1974.

[7] C. de Boor, *A Practical Guide to Splines*, Springer-Verlag, New York, 1978.

[8] C. de Boor, K. Höllig, and M. Sabin, *High accuracy geometric Hermite interpolation*, Comput. Aided Geom. Des., 4 (1987), pp. 269–278.

[9] G. Farin, *Curves and Surfaces for Computer Aided Geometric Design*, Academic Press, Boston, 1988.

[10] ——, *Curvature continuity and offsets for piecewise conics*, ACM Trans. Graphics, 8 (1989), pp. 89–99.

[11] R. T. Farouki, *The approximation of non-degenerate offset surfaces*, Comput. Aided Geom. Des., 3 (1986), pp. 15–43,

---

[5]Since the rational Bézier form is a de facto industry standard, we can also *transfer* offset profiles precisely between different CAD systems.

[12] ——, *Computing with barycentric polynomials*, Math. Intelligencer, 13 (1991), to appear.

[13] ——, *Hermite interpolation by Pythagorean–hodograph quintics*, in preparation.

[14] R. T. Farouki and C. A. Neff, *Analytic properties of plane offset curves*, Comput. Aided Geom. Des., 7 (1990), pp. 83–99.

[15] ——, *Algebraic properties of plane offset curves*, Comput. Aided Geom. Des., 7 (1990), pp. 101–127.

[16] R. T. Farouki and V. T. Rajan, *On the numerical condition of polynomials in Bernstein form*, Comput. Aided Geom. Des., 4 (1987) pp. 191–216.

[17] ——, *Algorithms for polynomials in Bernstein form*, Comput. Aided Geom. Des., 5 (1988), pp. 1–26.

[18] R. T. Farouki and T. Sakkalis, *Pythagorean hodographs*, IBM J. Res. Develop., 34 (1990), pp. 736–752.

[19] ——, *Real rational curves are not "unit speed,"* Comput. Aided Geom. Des., 8 (1991), pp. 151–157.

[20] H. Goldstein, *Classical Mechanics*, Second Edition, Addison-Wesley, Reading, MA, 1980.

[21] F. Gomes Teixeira, *Traité des Courbes Spéciales Remarquables Planes et Gauches* I,II,III, Chelsea, New York, reprint, 1907.

[22] T. N. T. Goodman and K. Unsworth, *Shape preserving interpolation by parametrically defined curves*, SIAM J. Numer. Anal., 25 (1988), pp. 1453–1465.

[23] T. N. T. Goodman and K. Unsworth, *Shape preserving interpolation by curvature continuous parametric curves*, Comput. Aided Geom. Des., 5 (1988), pp. 323–340.

[24] B. Guenter and R. Parent, *Computing the arc length of parametric curves*, IEEE Comput. Graphics Appl., 10 (1990), pp. 72–78.

[25] P. Henrici, *Elements of Numerical Analysis*, John Wiley, New York, 1964.

[26] J. Hoschek, *Offset curves in the plane*, Comput. Aided Des., 17 (1985), pp. 77–82.

[27] —— *Spline approximation of offset curves*, Comput. Aided Geom. Des., 5 (1988), pp. 33–40.

[28] J. Hoschek and N. Wissel, *Optimal approximate conversion of spline curves and spline approximation of offset curves*, Comput. Aided Des., 20 (1988), pp. 475–483.

[29] J. Hunter, *Number Theory*, Oliver and Boyd, Edinburgh, 1964.

[30] R. Klass, *An offset spline approximation for plane cubic splines*, Comput. Aided Des., 15 (1983), pp. 297–299.

[31] K. K. Kubota, *Pythagorean triples in unique factorization domains*, Amer. Math. Monthly, 79 (1972), pp. 503–505.

[32] J. D. Lawrence, *A Catalog of Special Plane Curves*, Dover, New York, 1972.

[33] R. R. Martin, *Principal patches for computational geometry*, Ph.D. Thesis, Cambridge University, 1982.

[34] B. Pham, *Offset approximation of uniform B-splines*, Comput. Aided Des., 20 (1988), pp. 471–474.

[35] G. Salmon, *A Treatise on the Higher Plane Curves*, Chelsea, New York, reprint, 1879.

[36] T. W. Sederberg and R. J. Meyers, *Loop detection in surface patch intersections*, Comput. Aided Geom. Des., 5 (1988), pp. 161–171.

[37] T. W. Sederberg and X. Wang, *Rational hodographs*, Comput. Aided Geom. Des., 4 (1987), pp. 333–335.

[38] R. J. Sharpe and R. W. Thorpe, *Numerical method for extracting an arc length parameterization from parametric curves*, Comput. Aided Des., 12 (1982), pp. 79–81.

[39] O. N. Stavroudis, *The Optics of Rays, Wavefronts, and Caustics*, Academic Press, New York, 1972.

[40] J. L. Synge and B. A. Griffith, *Principles of Mechanics*, Second Edition, McGraw-Hill, New York, 1949.

[41] W. Tiller and E. G. Hanson, *Offsets of two-dimensional profiles*, IEEE Comput. Graphics Appl., 4 (1984), pp. 36–46.

Self-Intersections and Offset Surfaces

Robert E. Barnhill, Todd M. Frost, and Scott N. Kersey

2.1. Introduction

Surface self-intersection is of interest in many applications, such as numerical control machining. For practical applications, it is desirable to incorporate an algorithm capable of supplying the solutions to the self-intersection of general parametric surfaces as it becomes cumbersome to handle different surface types as special cases. It is very important in many applications to have a robust surface self-intersection algorithm available to detect inappropriate solutions when they arise. In the case of numerical control machining, the cost of inappropriate tool selection resulting in surface gouging is extremely high.

The need for a surface self-intersection algorithm capable of handling general parametric surfaces is motivated by examining the description of self-intersection problems, the properties of a widely used complex parametric surface (an offset surface), and the algorithm requirements for an interactive computer graphics program within an engineering environment.

2.2. Problem Description

Parametric surfaces of a complex nature are often the basis for the problem of surface self-intersection and motivate the need for self-intersection algorithms which are not restricted to special classes of parametric surfaces. Construction of robust algorithms for computing self-intersections of general parametric surfaces is a very difficult problem. Houghton et al. [6], Barnhill et al. [3], and more recently Barnhill and Kersey [2] have approached the surface-surface intersection problem with generality in mind.

The solution to the self-intersection problem of a smooth parametric surface $\mathbf{r}(u,v)$ is defined as the set

$$\{\,(u_1, v_1, u_2, v_2) \mid \mathbf{r}(u_1, v_1) = \mathbf{r}(u_2, v_2) \text{ and } (u_1, v_1) \neq (u_2, v_2)\,\}.$$

Following Barnhill et al. [3], we classify these solutions as:

(i) empty,

(ii) a collection of points,

(iii) a collection of smooth curves,

(iv) a collection of smooth surfaces, or

(v) any combination of (ii), (iii), and (iv).

Surfaces which exhibit solutions of the form (i) or (iii) are considered to be in a "general" position. Surfaces not in a general position give rise to more difficult problems. Solutions of (ii) are in general hopeless for any nonexhaustive, very specialized algorithms. Solutions of (iv) can usually be detected using numerical methods and are handled with special consideration. The focus of the algorithm is to be able to handle solutions of surfaces in a "general" position.

2.3. Application: Offset Surfaces

Generation of navigation paths for numerical control machining is among the many applications in which self-intersection occurs. This application involves offset surfaces. Offset surfaces represent a complex class of potentially self-intersecting surfaces which are widely used. The complexity of an offset surface and its potentially pathological behavior has been examined in detail by Farouki [4]. The common definition of an offset surface $\mathbf{o}(u, v)$ to a smooth parametric progenitor surface $\mathbf{r}(u, v)$ can be written as

$$\mathbf{o}(u, v) = \mathbf{r}(u, v) + d\mathbf{n}(u, v),$$

where

$$\mathbf{n}(u, v) = \frac{\mathbf{r}_u \times \mathbf{r}_v}{\|\mathbf{r}_u \times \mathbf{r}_v\|}$$

is the unit surface normal and d is the offset distance.

The complexity of the offset surface can be observed from the unit surface normal $\mathbf{n}(u, v)$, which contains a cross product and, more significantly, division by a square root. Because of this complexity, offset surfaces are generally not within the same class as their progenitors (e.g., the offset of a polynomial surface is generally not polynomial). Systems capable of determining self-intersection within a particular class of surfaces might not be able to determine the self-intersection of their offsets if the intersection algorithm has been designed to exploit only the fundamental properties of their progenitor's class.

Various methods for approximation of parametric offset surfaces have been presented by Farouki [4], Hoschek, Schneider, and Wassum [5], and Patrikalakis and Prakash [9]. Approximation of an offset surface can allow the progenitor and offset to be represented within the same class of functions. However, if self-intersections occur in the offset surface, the approximation can become inaccurate in the region of self-intersection. For an accurate description of where self-intersections occur, it therefore becomes desirable to use the exact offset surface representation as the basis for the self-intersection problem, as noted by Aomura and Uehara [1].

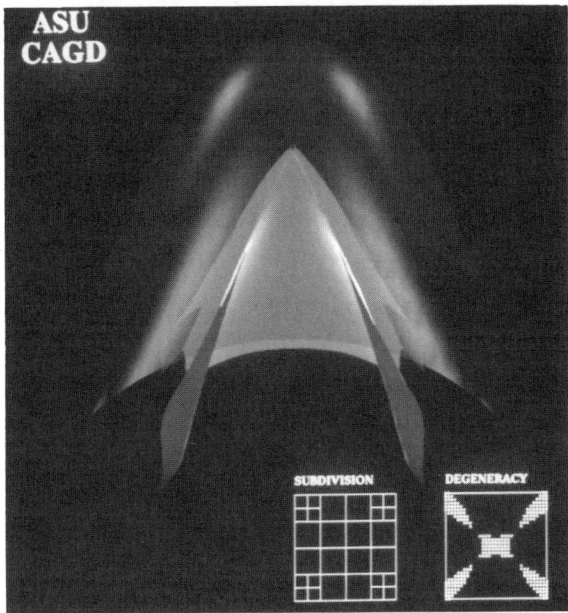

FIG. 2.1. *Progenitor (transparent) and degenerate offset surface.*

Self-intersection of an offset surface occurs when the offset distance is greater than the minimum *concave* radius of curvature of the progenitor surface. Generally, this results in loops in the offset surface which are bounded by curves of self-intersection [4]. This behavior can be visualized as trying to mill a detailed portion of a part with a spherical cutter whose radius is too large. A progenitor surface and its offset exhibiting such degenerate behavior are illustrated in Fig. 2.1. Self-intersection of an offset surface can also occur as a result of more general surface geometry which cannot be detected through local curvature analysis.

Indications of self-intersections occurring as a result of an offset distance which is too large can be given by performing curvature analysis on the progenitor surface. Numerical techniques in differential analysis may also be used to determine the maximum offset distance that can be used without inducing such self-intersection. However, self-intersection resulting from more general surface geometry must be detected with self-intersection algorithms. Additionally, in situations where a tool change or surface reformulation is impossible or inappropriate, self-intersection algorithms provide an accurate, detailed description of where self-intersections occur. These solutions can then be used to identify boundaries of inappropriate regions in the offset surface so that reformulation is no longer required; rather, portions of the initial offset surface definition are identified as invalid.

2.4. Algorithm Requirements

As part of an interactive computer graphics program in an enginering environment, a surface self-intersection algorithm should be accurate, robust, and efficient, as stated by Pratt and Geisow [11]. These measures should be balanced such that the algorithm produces reasonable results on realistic examples and the results are produced in a reasonable amount of time [3].

Tolerances are introduced that provide sufficient control over the algorithm such that a balance is met among the algorithm's requirements. Tolerances which can effectively be used to establish this balance include:

(1) same point tolerance (SPT),
(2) edge linearity and flatness measures, and
(3) step length based on curvature approximation.

SPT is used as a proximity measure for points in $\mathbb{R}^3$. Edge linearity refers to a linearity assessment of the boundary curves of a subpatch. For each subpatch, linearity is measured by examining the angles between corner tangent vectors in the direction of isoparametric curves.

Flatness refers to a planarity assessment of a subpatch. For each subpatch, flatness is measured by examining the angles between corner normal vectors and an interior normal vector. Step length is controlled in the marching scheme by examining an approximation to the curvature of the intersection curve.

2.5. Algorithm Description

The algorithm now described is based on the previously discussed problem and requirements specifications. The algorithm proceeds by obtaining starting points on intersection curves, marching along intersection curves, and sorting disjoint intersection approximation segments.

Start points are obtained through a surface subdivision process. The surface is adaptively subdivided until edge linearity and flatness tolerances are satisfied on all subpatches. When this has been accomplished, approximate bounding boxes are constructed for each subpatch. A set of approximate intersection points is then generated by computing the average of the corner points of conflicting, subpatch-bounding box pairs. These approximate intersection points are then relaxed onto intersection curves to establish a set of start points. Approximate intersection points are relaxed onto intersection curves using a Newton iteration to minimize

$$\|\mathbf{r}(u_1, v_1) - \mathbf{r}(u_2, v_2)\|$$

subject to the constraint

$$(u_1 - u_2)^2 + (v_1 - v_2)^2 > \epsilon > 0.$$

Marching proceeds by stepping along the approximate intersection curve's tangent vectors from starting and successive points along intersection curves. Step length is based on intersection curve arc length approximation. Tracing an individual intersection segment terminates when the segment being traced:

(1) meets the initial point of the same segment,

(2) meets a boundary, or

(3) meets another intersection segment.

Interference checks among the linear approximation segments indicate when one of the above termination conditions occur. The first termination condition indicates that the intersection segment is a closed loop. The second termination condition indicates that the intersection segment has crossed one of the domain boundaries and that the last approximate intersection point generated by the marching process must be relaxed onto the surface boundary as well as relaxed onto an intersection curve. The third termination condition indicates that tangent points (points on different intersection segments which share a common tangent plane) or branch points (points where multiple intersection segments meet) may exist. Various numerical methods are then used to determine if a tangent or branch point exists and to identify the location of the termination point. The marching process terminates when all start points have been exhausted.

Sorting connects and orders curve approximation segments generated from the marching process. Connectivity is determined by segments with common end points. Since approximation curve segments are generated in an ordered fashion using marching methods, sorting is not difficult to implement. However, some care must be taken to ensure that the various types of termination conditions which can occur are handled appropriately.

The following is a high level summary of the algorithm.

ALGORITHM.

A. Generate start points.

1. Subdivide surface until subpatches are "flat."
2. Generate bounding boxes of subpatches.
3. Average conflicting subpatch pair corner vertices.
4. Relax points to an intersection.

B. March along intersection curves.

1. Get step vector.
2. Relax point to an intersection.
3. Check for conflicting intersections.
4. Relax points onto boundaries if necessary.
5. Compute tangent or branch points if necessary.
6. Continue for all start points.

C. Sort disjoint intersection approximation segments.

The generality of this method is manifested in its ability to solve self-intersection problems involving arbitrary, smooth parametric surfaces given position and partial derivative evaluators. The algorithm has been implemented such that parametric surfaces based on rectangular and triangular domains can be processed.

FIG. 2.2. *Simple bicubic patch.*

Figures 2.2–2.6 illustrate various self-intersection problems and their solutions. For a more detailed description of the algorithm and implementation details, see Barnhill and Kersey [2].

2.6. Further Research

A recent development by Kriezis and Patrikalakis [7],Patrikalakis, Kriezis, and Prakash [10], and Patrikalakis [8] involves subdividing thesurface through identification of *significant* points of the intersection curve. Significant points of an intersection curve are defined as border, turning, and singular points. Border points occur when at least one of the parametric variables takes on a value equal to the border of the parametric domain. Turning points occur when the tangent vector of the pre-image of the intersection curve is parallel to one of the parametric boundaries. Singular points occur when the surface normals are collinear at a point of intersection. An intersection curve segment is monotonic in any of the subpatches formed by specialized subdivision of the domain based on these significant points. The idea behind this type of subdivision is that the marching method can be applied more judiciously, the connectivity of an approximation can be determined more reliably, and the solution may be more vigorously examined.

While this domain subdivision technique has been effectively used for rational B-spline surfaces, the efficient extension of this method to complex general parametric surfaces such as the offset surface requires further research. This results from the exploitation of the basic properties of the rational B-spline in various steps used in the algorithm. An excellent description of the outstanding problems in surface-surface intersection is given in Patrikalakis [8].

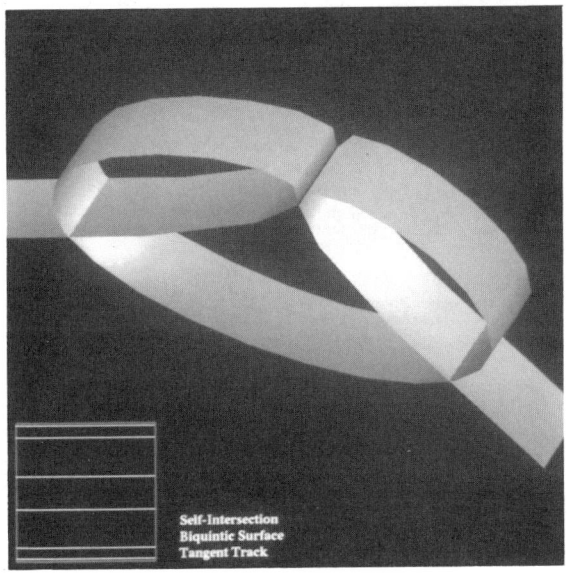

FIG. 2.3. *Complex biquintic patch.*

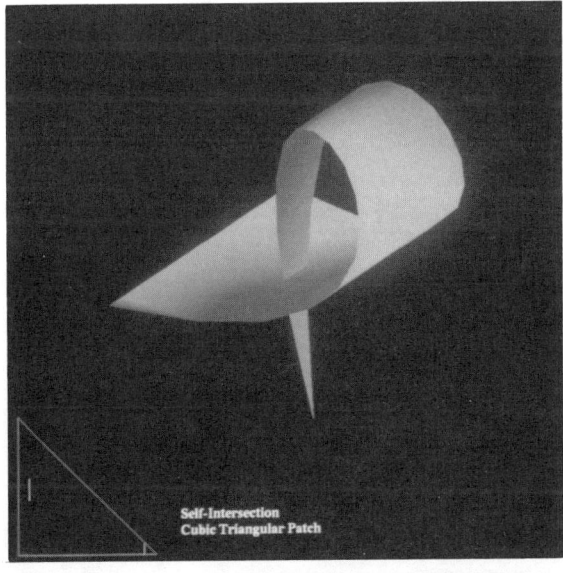

FIG. 2.4. *Simple cubic triangular patch.*

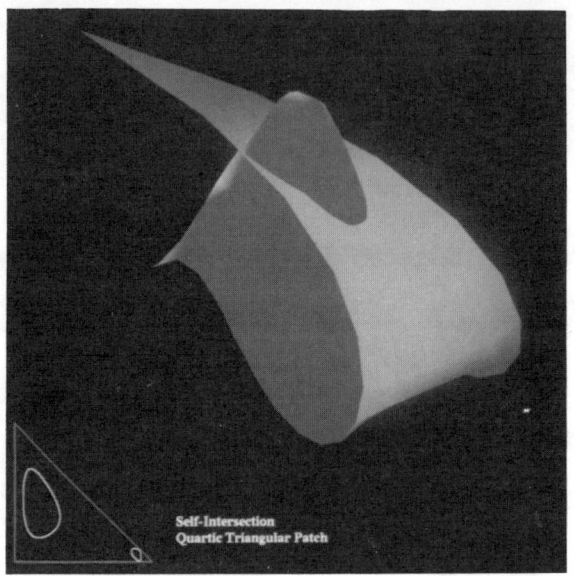

FIG. 2.5.   *Complex quartic triangular patch.*

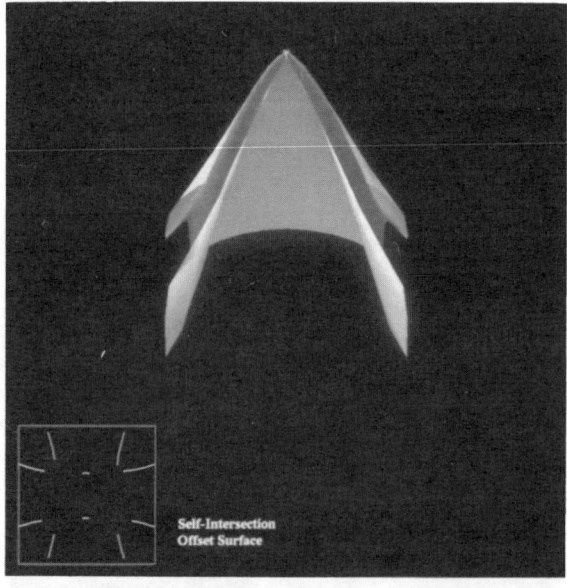

FIG. 2.6.   *Degenerate offset surface.*

2.7. Conclusion

The key to developing a self-intersection algorithm which is transportable to many systems is the ability of the algorithm to handle general parametric surfaces. This is illustrated by the complexity and widely varying classes of parametric surfaces requiring self-intersection testing. The parametric offset surface is one particular example of such a complex surface. The marching method described in this paper is an example of a self-intersection algorithm which uses only surface position and partial derivative evaluators and does not require any specialized information.

Acknowledgments

The authors would like to acknowledge Wayne Woodland for his assistance in generating photographs. This research was supported by National Science Foundation grant DMC-8807747 and Department of Energy grant DEFG0287ER25041 to Arizona State University, Tempe, AZ.

References

[1] S. Aomura and T. Uehara, *Self-intersection of an offset surface*, Comput. Aided Des., 22 (1990), pp. 417–422.

[2] R.E. Barnhill and S.N. Kersey, *A marching method for parametric surface-surface intersection*, Comput. Aided Geom. Des., 7 (1990), pp. 257–280.

[3] R.E. Barnhill, G. Farin, M. Jordan, and B.R. Piper, *Surface-surface intersection*, Comput. Aided Geom. Des., 4 (1987), pp. 3–16.

[4] R.T. Farouki, *The approximation of non-degenerate offset surfaces*, Comput. Aided Geom. Des., 3 (1986), pp. 15–43.

[5] J. Hoschek, F.-J. Schneider, and P. Wassum, *Optimal approximate conversion of spline surfaces*, Comput. Aided Geom. Des., 6 (1989), pp. 293–306.

[6] E.G. Houghton, R.F. Emnett, J.D. Factor, and C.L. Sabharwal, *Implementation of a divide-and-conquer method for intersection of parametric surfaces*, Comput. Aided Geom. Des., 2 (1985), pp. 173–183.

[7] G.A. Kriezis and N.M. Patrikalakis, *Interrogation of rational polynomial surface intersections*, Design Laboratory Memorandum 89-9, Department of Ocean Engineering, Massachusetts Institute of Technology, Cambridge, MA, October 1989.

[8] N.M. Patrikalakis, *Interrogation of surface intersections*, in Geometry Processing for Design and Manufacturing, R.E. Barnhill, ed., Society for Industrial and Applied Mathematics, Philadelphia, PA, 1991, pp. 161–185.

[9] N.M. Patrikalakis and P.V. Prakash, *Free form plate modeling using offset surfaces*, in Computers in Offshore and Arctic Engineering – 1987, J.S. Chung and D. Angelides, eds., American Society of Mechanical Engineers, New York, 1987, pp. 37–44.

[10] N.M. Patrikalakis, G.A. Kriezis, and P.V. Prakash, *An investigation on surface-to-surface intersection problems*, 16th Annual National Science Foundation Conference on Manufacturing Systems Research, Tempe, AZ, 1990.

[11] M.J. Pratt and A.D. Geisow, *Surface/surface intersection problems*, in The Mathematics of Surfaces, J.A. Gregory, ed., Oxford University Press, Oxford, 1986.

Bibliography

[1] R.E. Barnhill, *Geometry processing: curvature analysis and surface-surface intersection*, in Mathematical Methods in Computer Aided Geometric Design, T. Lyche and L.L. Schumaker, eds., Academic Press, New York, 1989, pp. 51–60.

[2] Y.J. Chen and B. Ravani, *Offset surface generation and contouring in computer aided design*, J. Mech. Trans. Auto. Des., Trans. ASME, 109 (1987), pp. 133–142.

[3] R.T. Farouki, *Exact offset procedures for simple solids*, Comput. Aided Geom. Des., 2 (1985), pp. 257–279.

[4] J. Hoschek, *Offset curves in the plane*, Comput. Aided Des., 17 (1985), pp. 77–82.

[5] R. Klass, *An offset spline approximation for plane cubic splines*, Comput. Aided Des., 15 (1983), pp. 297–299.

[6] R.P. Markot and R.L. Magedson, *Solutions of tangential surface and curve intersections*, Comput. Aided Des., 21 (1989), pp. 421–429.

[7] G. Mullenheim, *Convergence of a surface/surface intersection algorithm*, Comput. Aided Geom. Des., 7 (1990), pp. 415–423.

[8] B. Pham, *Offset approximation of uniform B-splines*, Comput. Aided Des., 20 (1988), pp. 471–474.

[9] W. Tiller and E.G. Hanson, *Offsets of two-dimensional profiles*, IEEE Comput. Graphics Appl., 4 (1984), pp. 36–46.

Approximate Spline Conversion for Integral and Rational Bézier and B-Spline Surfaces

Josef Hoschek and Franz-Josef Schneider

3.1. Introduction

A great majority of computer aided design (CAD) systems for free-form curve and surface modeling use parametric representations; nowadays also rational representations are introduced because of their larger degree of freedom and their exact representation of conics. The new industrial interface STEP deals also with rational surfaces. The representation schemes used within these systems nevertheless differ a lot with regard to the types of polynomial bases and the maximum polynomial degrees provided. Bernstein–Bézier, B-spline, and monomial basis functions are frequently used in different systems. Polynomial degrees vary between 3 and about 20 as the available upper bound.

With the availability of a fast growing variety of modeling systems, the demand has risen to exchange curve and surface descriptions between one CAD system and another. Particularly in large industrial organizations where a multitude of modeling systems exists, there is a need for communications of geometry descriptions between different departments within the firm or with outside manufacturers and subcontractors. Therefore, the need for methods of conversion of surface representations was recognized at an early stage. Conversion from one polynomial base to another can be achieved by direct matrix multiplication whenever the number and degrees of polynomial terms in both representations are equal or the degree of the polynomials should be elevated [12], [10], [8].

If two systems do not allow for the same maximum polynomial degrees, then approximate conversions of high degree into low order functions (reducing combined with splitting spline segments) and perhaps vice versa (elevating and merging spline segments) are inevitable. This causes approximation errors which must be minimized. Dannenberg and Nowacki [7] have introduced a first approach which uses an error estimate due to de Boor [6] and an application of this estimate due to Hölzle [13] to evaluate a new segmentation of the spline

---

*Portions of this paper appeared in *Computer Aided Design*, 1990, vol. 22, pp. 580–590. By permission of the publishers, ©Butterworth Heinemann Ltd.

curve. This method is implemented in the German VDA Software sponsored by the German Association of Automobile Manufacturers. Unfortunately, a large number of new patches often arises using this method (see [7]). Therefore, new developments were worked out. Chebyshev polynomials are used for the conversion process [24]; Bézier curves [22], [29] and B-spline curves [3] are converted. The last method was extended to tensor product, B-spline surfaces [3]. The number of the obtained patches is similar to the number in the method developed in [7] (compare [4]).

In the present paper we give an overview of the effective approximate conversion methods from rational or integral B-spline curves or B-spline surfaces, respectively; rational or integral Bézier curves or Bézier surfaces to integral Bézier curves or Bézier surfaces, which are developed in [15], [21], [20], [18]. The introduced methods can be extended to spline representation of offset curves and offset surfaces [16], [21], [20].

3.2. Conversion of Curves

3.2.1. Geometric Continuity.
The key idea of the proposed method is to use parameterization as a design parameter. The shape of an approximation curve or surface of a set of points will be changed if the parameter values of the points are changed during the approximation process. If we want to change parameterization during the design process, we must use spline conditions which are invariant to parameterization. We can use osculating conditions which are well known in differential geometry [11], [5] as conditions of *contact of order k* of two curves or surfaces.

Two curves $\mathbf{X}$ and $\mathbf{Y}$ have G^k continuity if the following conditions hold at a common point of $\mathbf{X}$ and $\mathbf{Y}$:

$$k = 1 : \mathbf{X}^{\mathrm{i}} = \lambda_1 \mathbf{Y}^{\mathrm{i}},$$
$$k = 2 : \mathbf{X}^{\mathrm{ii}} = \lambda_1^2 \mathbf{Y}^{\mathrm{ii}} + \lambda_2 \mathbf{Y}^{\mathrm{i}} \quad (\text{and conditions for } k = 1),$$
$$k = 3 : \mathbf{X}^{\mathrm{iii}} = \lambda_1^3 \mathbf{Y}^{\mathrm{iii}} + 3\lambda_1\lambda_2 \mathbf{Y}^{\mathrm{ii}} + \lambda_3 \mathbf{Y}^{\mathrm{i}} \quad (\text{and conditions for } k = 1, 2),$$
$$k = 4 : \mathbf{X}^{\mathrm{iv}} = \lambda_1^4 \mathbf{Y}^{\mathrm{iv}} + 6\lambda_1^2\lambda_2 \mathbf{Y}^{\mathrm{iii}} + (3\lambda_2^2 + 4\lambda_1\lambda_3)\mathbf{Y}^{\mathrm{ii}} + \lambda_4 \mathbf{Y}^{\mathrm{i}}$$
$$(\text{and conditions for } k = 1, 2, 3)$$

(3.1)

with arbitrarily chosen parameters λ_j. These conditions can be developed by using an osculating algebraic curve of degree k or out of the first k terms of the Taylor expansion [12].

The conditions in (3.1) can be described in the following recursion formula:

$$\mathbf{X}^{(n)}(0) = \sum_{l=1}^{n} \omega_{nl} \mathbf{Y}^{(l)}(0) \quad (\text{with } \omega_{00} = 1, \omega_{11} = \lambda_1, \omega_{lo} = 0; \omega_{lk} = 0 \text{ for } l < k)$$

and the recursion for the ω_{ki}

$$\omega_{ki} = \lambda_1 \omega_{k-1,i-1} + E\omega_{k-1,i}.$$

The *shifting operator* E has the properties

$$E\lambda_j = \lambda_{j+1}, \qquad E(\lambda_j\lambda_k) = \lambda_{j+1}\lambda_k + \lambda_j\lambda_{k+1}.$$

3.2.2. Approximate Degree Reduction. The given spline curve may have a B-spline representation of (arbitrary) order k

$$(3.2) \qquad \mathbf{X}(t) = \frac{\sum_{i=0}^{p}\beta_i\mathbf{d}_iN_{ik}(t)}{\sum_{i=0}^{p}\beta_iN_{ik}(t)}.$$

The basis function may be defined over a uniform knot vector while the first and the last knot values have multiplicity k (see [12]). The $\mathbf{d}_i$ are the de Boor points, the β_i the weights. We can assume that the parameter is running through $t \in [0,a]$; then we have $s = p - k + 2$ segments if only simple knots are used in the interior of the knot vector. If each knot in the knot vector has multiplicity k, the B-spline representation (3.2) is transformed in a Bézier representation of the same curve.

In general, the boundary points of the B-spline segments are arbitrarily given. They are established by experience by the patch designer. As our approximation process depends on the boundary points of the segments, first we will introduce a new approach to find natural, geometric-oriented boundary points which split the given curve into a set of Bézier curves or B-spline curves.

Because of the variation diminishing property [8], [12], [26], a *generic* (plane) cubic Bézier curve in the region used in practice has no more than one (interior) minimum of curvature. As demonstrated in [26], a cubic Bézier curve can have more than one minimum of curvature in general, but these general cases should not be applied in practice. Figure 3.1 gives an overview on the distribution of minima of curvature of generic cubic Bézier curves. If we consider a quintic curve, we get no more than three minima of curvature in the generic case. Therefore, a cubic curve (or a quintic one) cannot be expected to approximate a curve with more than one (more than three) minimum (minima) of curvature!

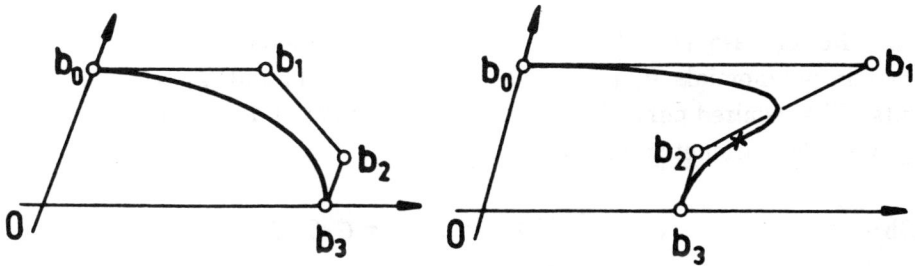

FIG. 3.1. *Generic cubic Bézier curves and their inflection points.*

If a given curve has more than one minima of curvature in the cubic case (more than three in the quintic case), the curve must be subdivided in more than one segment if optimal approximation results are wanted.

To determine new boundary points of a given system of spline segments, we first discretize the B-spline curve with the help of suitably chosen parametric values t_i. We find a minimum of curvature when the following condition holds:

$$\kappa(t_{i-1}) - \kappa(t_i) < 0 \wedge \kappa(t_{i+1}) - \kappa(t_i) > 0.$$

Now in the cubic case the curve is split midway between each minimum; in the quintic case the curve is split midway between the third and fourth minima and so on. After the splitting procedure, the given B-spline curve is split into p_1 Bézier or B-spline curves. If a required error estimate ϵ_0 is not obtained through the approximation process, the segments with approximation error larger then ϵ_0 must be subdivided additionally. The points with maximal deviation are used as new boundary points.

Whether we use the B-spline or the Bézier representation for the new curve segments depends on the position of the new boundary points on the knot vector. If two parameter values of the new boundary points are within an interval or on a knot of the given knot vector, we introduce Bézier curves; otherwise we use B-spline curves (see Fig. 3.2).

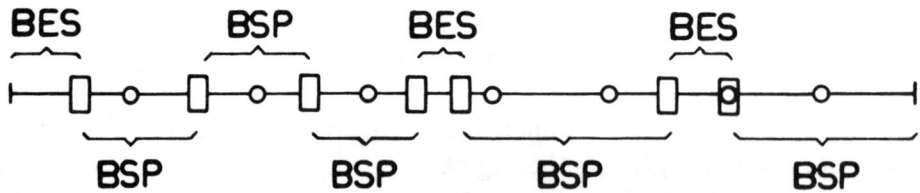

FIG. 3.2.   *Splitting in B-spline* (BSP) *or Bézier curves* (BES) (o *knots,* □ *new boundaries*).

Now each of these curve segments will be converted into Bézier curves. The given curve (spline segment) **X** may have the parametric representation

$$(3.3a) \qquad \mathbf{X} = \sum_{i=0}^{n} \mathbf{V}_i B_i^n(t), \qquad t \in [0,1],$$

in the Bézier case (the B-spline representation follows analogously) with Bernstein polynomials $B_i^m(t)$ of degree n ($n = k - 1$) and $\mathbf{V}_i$ as given Bézier points. The required curve **Y** may be a Bézier curve of degree m ($m < n$) and may have the parametric representation

$$(3.3b) \qquad \mathbf{Y} = \sum_{i=0}^{m} \mathbf{W}_i B_i^m(\tau), \qquad \tau \in [0,1),$$

with unknown Bézier points $\mathbf{W}_i$. m is bounded by $m \geq 2k + 1$ with k as the order of the continuity conditions. We transform conditions (3.1) into *boundary conditions* for the points $\mathbf{X}(0) = \mathbf{Y}(0)$, $\mathbf{X}(1) = \mathbf{Y}(1)$, and obtain after some calculations the following conditions for the unknown Bézier points $\mathbf{W}_i$ (see [15], [21]):

$k = 1$:

(3.4a)
$$\mathbf{W}_0 = \mathbf{V}_0, \qquad \mathbf{W}_m = \mathbf{V}_n,$$
$$\mathbf{W}_1 = \mathbf{V}_0 + (\mathbf{V}_1 - \mathbf{V}_0)\lambda_1, \qquad \mathbf{W}_{m-1} = \mathbf{V}_n + (\mathbf{V}_{n-1} - \mathbf{V}_n)\mu_1,$$

$k = 2$:

(3.4b)
$$\mathbf{W}_2 = \mathbf{V}_0 + (\mathbf{V}_2 - \mathbf{V}_1)\lambda_1^2\omega_1 + (\mathbf{V}_1 - \mathbf{V}_0)\lambda_2,$$
$$\mathbf{W}_{m-2} = \mathbf{V}_n + (\mathbf{V}_{n-2} - \mathbf{V}_{n-1})\mu_1^2\omega_1 + (\mathbf{V}_{n-1} - \mathbf{V}_n)\mu_2,$$

(with (3.4a)),

$k = 3$:

(3.4c)
$$\mathbf{W}_3 = \mathbf{V}_0 + (\mathbf{V}_3 - \mathbf{V}_2)\lambda_1^3\omega_2$$
$$+ (\mathbf{V}_2 - \mathbf{V}_1)\left[\lambda_1^3\left(-2\omega_2 + 3\omega_1^2\frac{m-1}{m-2}\right)\right.$$
$$\left. + \lambda_1^2\omega_1\left(3 - 6\frac{m-1}{m-2}\right) + 3\lambda_1\lambda_2\omega_1\frac{m-1}{m-2}\right] + (\mathbf{V}_1 - \mathbf{V}_0)\lambda_3$$

(with (3.4a,b)),

and analogous conditions for $\mathbf{W}_{m-3}$ and $\mathbf{W}_{m-4}$ (the λ_i have to be replaced by μ_i). The quantities λ_i, μ_i are arbitrarily chosen new parameters (which are abbreviations for the coefficients generated during calculation).

Further abbreviations are introduced:

(3.5)
$$\omega_1 = \frac{m(n-1)}{n(m-1)}, \qquad \omega_2 = \frac{m^2(n-1)(n-2)}{n^2(m-1)(m-2)}.$$

3.2.3. Approximation Error for Degree Reducing. Our goal is to approximate the given Bézier curve $\mathbf{X}$ by a Bézier curve $\mathbf{Y}$ optimally, where *optimally* means *minimizing the square error sum*. The position error will be measured at $s + 1$ points $\mathbf{P}_i$ of the given Bézier curve $\mathbf{X}$ of degree $n(s > n)$ with the (chordal) parameter values t_i,

(3.6)
$$\mathbf{P}_i = \mathbf{X}(t_i), \qquad i = 0(1)s.$$

If we insert these parameter values into the required Bézier curve $\mathbf{Y}$, we obtain as error vectors

$$\boldsymbol{\delta}_i = \mathbf{P}_i - \mathbf{Y}(t_i)$$

and as square error sum

$$\delta = \sum_{i=0}^{s} \boldsymbol{\delta}_i^2.$$

For the different continuity conditions the error vectors are determined by (with (3.3a–c), (3.4a–c), (3.5))

$k = 1$ and $m = 3$:

(3.7a)
$$\delta_i = \mathbf{R}_i - (\mathbf{V}_1 - \mathbf{V}_0)\lambda_1 B_1^3(t_i) - (\mathbf{V}_{n-1} - \mathbf{V}_n)\mu_1 B_2^3(t_i) \quad \text{with}$$
$$\mathbf{R}_i = \mathbf{P}_i - \mathbf{V}_0(B_0^3(t_i) + B_1^3(t_i)) - \mathbf{V}_n(B_2^3(t_i) + B_3^3(t_i)).$$

$k = 2$ and $m = 5$:

$$\delta_i = \mathbf{R}_i - (\mathbf{V}_1 - \mathbf{V}_0)(\lambda_1 B_1^5(t_i) + \lambda_2 B_2^5(t_i)) - (\mathbf{V}_2 - \mathbf{V}_1)\lambda_1^5\omega_1 B_2^5(t_i)$$
$$-(\mathbf{V}_{n-1} - \mathbf{V}_n)(\mu_1 B_4^5(t_i) + \mu_2 B_3^5(t_i))$$

(3.7b)
$$-(\mathbf{V}_{n-2} - \mathbf{V}_{n-1})\mu_1^2\omega_1 B_3^5(t_i) \quad \text{with}$$
$$\mathbf{R}_i = \mathbf{P}_i - \mathbf{V}_0(B_0^5(t_i) + B_1^5(t_i) + B_2^5(t_i))$$
$$-\mathbf{V}_n(B_3^5(t_i) + B_4^5(t_i) + B_5^5(t_i)).$$

We obtain similar conditions for $k = 3$ and $m = 7$, $k = 4$ and $m = 9$ (see [21]). In (3.7a,b) only the case $m = 2k + 1$ is considered. If $m > 2k + 1$, the undetermined inner Bézier points $\mathbf{W}_j$ are additional unknown variables.

3.2.4. Optimization Algorithm. For $k = 1$ we obtain a *linear* error sum (from (3.7a)), so we can use the least square method for minimizing the total error. The minimum of δ is determined by the conditions

(3.8)
$$\frac{\partial\delta}{\partial\lambda_1} = 0, \qquad \frac{\partial\delta}{\partial\mu_1} = 0.$$

These conditions lead for $k = 1$ to a linear system for λ_1, μ_1. For $k > 1$ we obtain *a nonlinear* error sum, so we have to use a *nonlinear optimization procedure*. We can use a numerical optimization algorithm which approximates the gradients of the objective function and leads to a minimum.

The obtained square error sum depends on parameterization; hence the error vectors δ_i are not orthogonal to the approximation curve $\mathbf{Y}$. To obtain error vectors δ_i approximately normal to the approximation curve $\mathbf{Y}(t)$, a *parameter correction* is used. To transfer the (oblique) error vectors referred to in (3.7a,b) into (approximately) orthogonal error vectors, we have to change parameterization of the points $\mathbf{P}_i$ with help of a suitable method. We replace the approximation curve $\mathbf{Y}(t)$ at each point $\mathbf{Y}(t_i)$ by the tangent T_i and choose the distance between the point $\mathbf{Y}(t)$ and the foot of the perpendicular as an approximation of the parameter correction (normed by a factor ω which can be an estimate of the arc length of the approximation curve $\mathbf{Y}(t)$) (see Fig. 3.3). With the help of this consideration, a first approximation of an optimal parameter value can be evaluated by (see also [15], [17], [12])

(3.9a)
$$t_i^* = t_i + \frac{\Delta c_i}{\omega}.$$

The convergence is faster if we use the parameter correction introduced in [27]. The correction is evaluated by minimizing the error $(\mathbf{P}_i - \mathbf{Y}(t_i + \Delta t_i))^2$. With the help of the Taylor expansion, we obtain

$$(3.9b) \qquad \Delta t_i = \frac{(\mathbf{P}_i - \mathbf{Y}(t_i)) \cdot \mathbf{Y}^{\mathrm{i}}(t_i)}{(\mathbf{P}_i - \mathbf{Y}(t_i)) \cdot \mathbf{Y}^{\mathrm{ii}}(t_i) - \mathbf{Y}^{\mathrm{i}}(t_i)^2} \quad \text{or} \quad t_i^* = t_i + \Delta t_i.$$

Now we insert these parameter values t_i^* in (3.6), repeat the described minimization procedures, and obtain a corrected approximation curve $\mathbf{Y}^*$. If we repeat the whole procedure several times, we will get error vectors $\boldsymbol{\delta}_i$ which converge to the normals of the approximation curve. Thus a parameterization is obtained which leads to the minimization of the shortest distances between the given points and the approximation curve. The obtained approximation curve is the best with respect to the (Euclidean) distance norm.

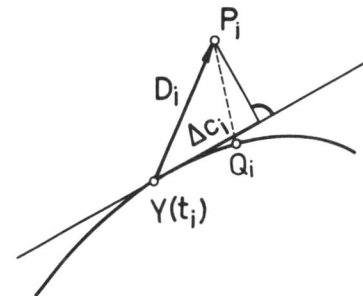

FIG. 3.3. *Correction of the parameterization of the approximation curve* $\mathbf{Y}(t)$.

The degree reduction is finished if the maximum error $\epsilon_1 = \max|\boldsymbol{\delta}_i| < \epsilon_0$ with ϵ_0 as given error tolerance. If the error ϵ_1 of one of these segments exceeds the given error tolerance ϵ_0, its segment will be split again at the parameter value with largest approximation error. The *de Casteljau algorithm* is used for splitting the spline segments. Figure 3.4 contains a diagram of the curve approximation algorithm.

Figure 3.5a contains a given Bézier curve of degree 19 and the approximating Bézier spline curve of degree $m = 5$ with geometric continuity of order $k = 2$ which has four segments. Additionally, the effect of the parameter correction is demonstrated. The Bézier points of the given curve segments are denoted by crosses; the Bézier points of the approximating curves are marked by boxes. Figure 3.5a shows the given and the approximating segments without parameter correction; in Fig. 3.5b the parameter correction is used for approximation. Both curves, the given one and the approximating curve segments, are drawn one above the other.

Remark. The obtained Bézier spline curves have geometric continuity and can therefore be interpreted as β-spline curves [2], [12].

3.3. Conversion of Rational B-Spline Surfaces

Now we will extend the method from §3.2 to tensor product surfaces. As we will use parameter correction, we have to introduce parameter invariant spline condition.

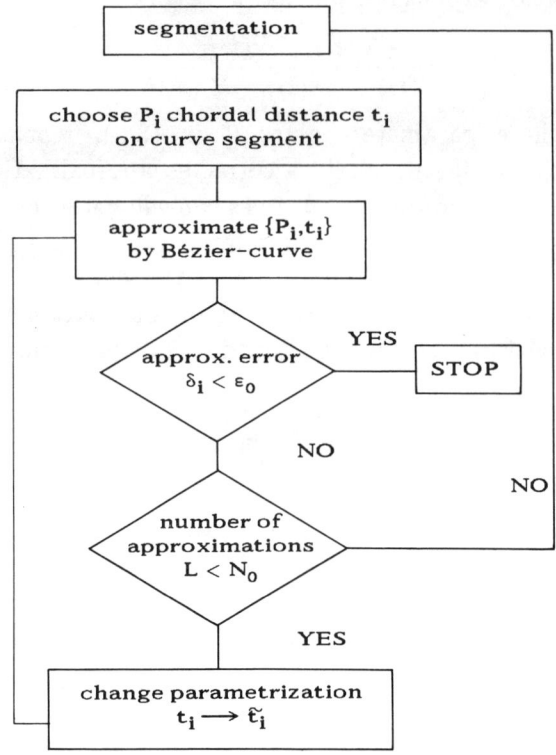

FIG. 3.4. *Algorithm of curve approximation with parameter correction.*

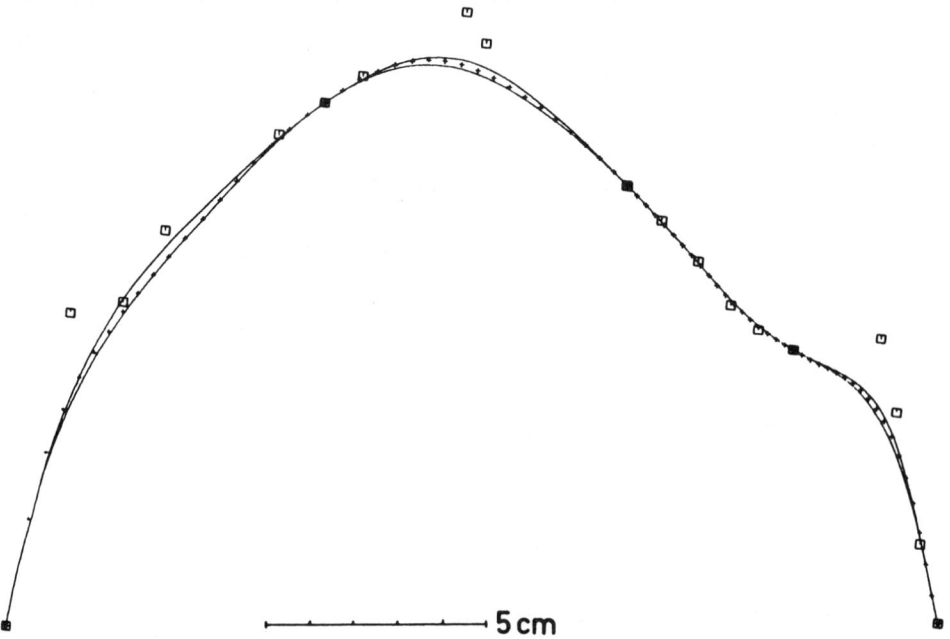

FIG. 3.5a. *Bézier curve of degree 19 and four approximating Bézier spline segments of degree 5 without parameter correction.*

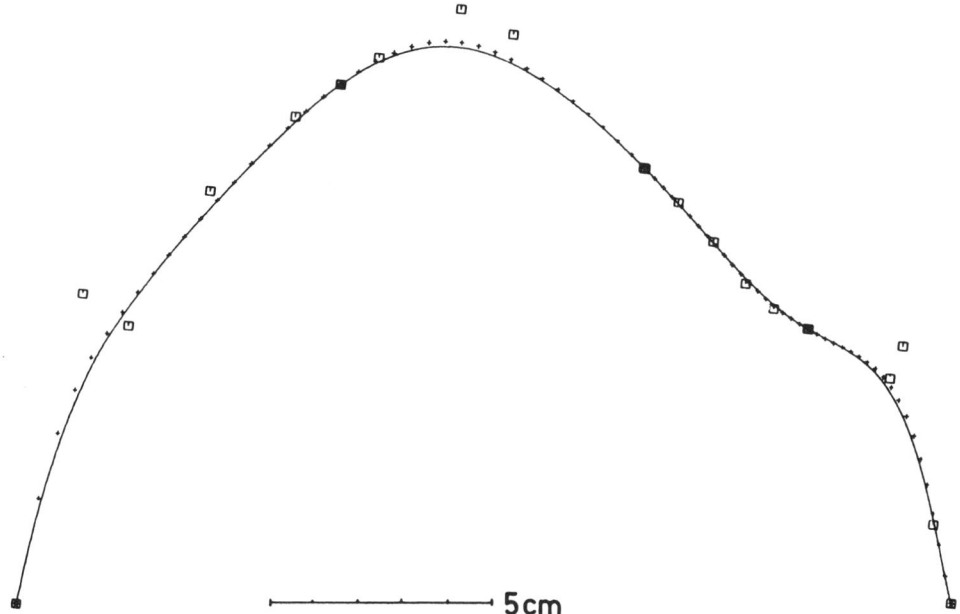

FIG. 3.5b. *Bézier spline curve out of Fig. 3.5a after parameter correction.*

3.3.1. Contact of Order k. Two surfaces $\mathbf{X}$ and $\mathbf{Y}$ have contact of order k if the following conditions hold at a common point of $\mathbf{X}$ and $\mathbf{Y}$ (a_{ik}, b_{ik} arbitrary factors) [5]:

$k = 1$:

$$(3.10) \qquad \begin{pmatrix} \mathbf{Y}_u \\ \mathbf{Y}_v \end{pmatrix} = \begin{pmatrix} a_{10} & b_{10} \\ a_{01} & b_{01} \end{pmatrix} \begin{pmatrix} \mathbf{X}_u \\ \mathbf{X}_v \end{pmatrix},$$

$k = 2$:

$$\begin{pmatrix} \mathbf{Y}_{uu} \\ \mathbf{Y}_{uv} \\ \mathbf{Y}_{vv} \end{pmatrix} = \begin{pmatrix} a_{20} & b_{20} \\ a_{11} & b_{11} \\ a_{02} & b_{02} \end{pmatrix} \begin{pmatrix} \mathbf{X}_u \\ \mathbf{X}_v \end{pmatrix} + \begin{pmatrix} a_{10}^2 & 2a_{10}b_{10} & b_{10}^2 \\ a_{10}a_{01} & a_{10}b_{01} + a_{01}b_{10} & b_{10}b_{01} \\ a_{02}^2 & 2a_{01}b_{01} & b_{01}^2 \end{pmatrix} \begin{pmatrix} \mathbf{X}_{uu} \\ \mathbf{X}_{uv} \\ \mathbf{X}_{vv} \end{pmatrix},$$

(3.11)

(and conditions for $k = 1$)

We can give the following geometric interpretation of these conditions. Surfaces with contact of order 1 in a common point $\mathbf{P}$ have a common tangent plane in $\mathbf{P}$. Surfaces with contact of order 2 in a common point $\mathbf{P}$ have the same Dupinian indicatrix in $\mathbf{P}$ and a common tangent plane.

3.3.2. Determination of Generic Boundary Curves. The goal is to develop an approximation procedure for the conversion of a given rational B-spline surface $\mathbf{X}(u, v)$ of (arbitrary) order (n, m) (see also [18]):

$$(3.12) \qquad \mathbf{X}(u, v) = \frac{\sum_{i=0}^{p} \sum_{k=0}^{q} \beta_{ik} \mathbf{d}_{ik} N_{in}(u) N_{km}(v)}{\sum_{i=0}^{p} \sum_{k=0}^{q} \beta_{ik} N_{in}(u) N_{km}(v)}.$$

The basis functions may be defined over nonuniform knot vectors while the first and last knot values have multiplicity n, m, respectively. $\mathbf{d}_{ik}$ are the de Boor points, β_{ik} the weights [12]. We can assume that the parameters in (3.12) are running through $u \in [0, a]$, $v \in [0, b]$. We have in u direction $s_1 := p - n + 2$ segments and in v direction $s_2 := q - m + 2$ segments. Thus the total number of B-spline segments is $s := s_1 s_2$. If all weights β_{ik} in (3.12) are equal to 1, the rational B-spline surface changes to an integral B-spline surface. Thus our developed procedure can also be used for integral B-spline surface. If $p = n - 1$, $q = m - 1$, (3.12) changes in the Bézier representation of a surface patch.

The **given** rational B-spline surfaces (3.12) are converted into a bicubic (or biquintic) set of integral Bézier patches

$$(3.13) \qquad \mathbf{Y}_{\rho\sigma}(u, v) = \sum_{i=0}^{3} \sum_{k=0}^{3} \mathbf{W}_{\rho\sigma,ik} B_i^3(u) B_k^3(v)$$

with (ρ, σ) as an index of the converted patches ($\rho = 1(1)\bar{\rho}$, $\sigma = 1(1)\bar{\sigma}$). The converted set of patches may have (in general) new boundary curves $u = 0$, $u = u_\rho$, $v = 0$, $v = v_\sigma$, with $u_\rho \in [0, a]$, $u_{\bar{\rho}} = a$, $v_\sigma \in [0, b]$, $v_{\bar{\sigma}} = b$, which in general do not coincide with the boundary curves of the given set. The total number of the required set may be $\bar{s} = \bar{\rho} \cdot \bar{\sigma}$ with $s \gtrless \bar{s}$.

In general the boundary curves of the B-spline segments are arbitrarily given. They are established by experience by the patch designer. While our approximation process depends on the boundary curves of the segments, we will first introduce a new approach to find natural, geometrically oriented boundary curves which split the given B-spline surface into a set of Bézier or B-spline segments.

The whole conversion process can be subdivided into following steps:

(i) Determine new geometric oriented boundary curves;

(ii) Approximate the new boundary curves by method developed in §3.2;

(iii) Approximate the interior of the new patches;

(iv) Approximate the curves of the trimmed surfaces.

With the same generic arguments introduced in §3.2.2 for the splitting of B-spline curves, we can develop a segmentation strategy which leads to new patch boundary curves (instead of boundary points as in §3.2.2). The goal of our segmentation strategy is twofold: (1) to shift the minima of curvature of a parametric line to different patches, and (2) to construct a minimal number of patches.

The segmentation runs with several iterative steps. We demonstrate the procedure for the approximation by bicubic surfaces and discretization of the given set in the u direction.

1. We discretize the parametric domain with respect to (suitably chosen) parametric differences Δu, Δv. This leads to new parametric lines $u = r \cdot \Delta u$, $v = s \cdot \Delta v$ ($r, s = 0, 1, 2 \cdots$).

2. Beginning at $u = 0$, we determine the two first minima of curvature on each discretized parametric line $v = $ const (see Fig. 3.6).

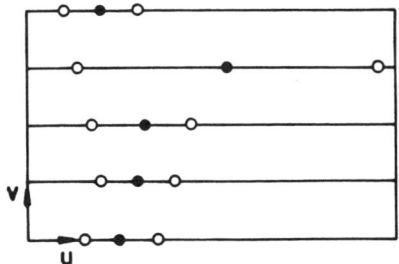

FIG. 3.6. *Local segmentation points in the parametric domain.*

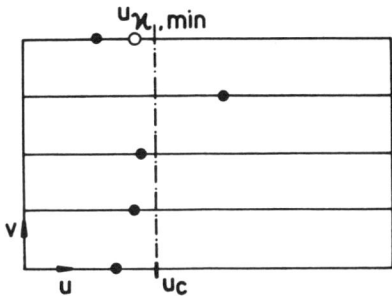

FIG. 3.7. *First segmentation line of a patch.*

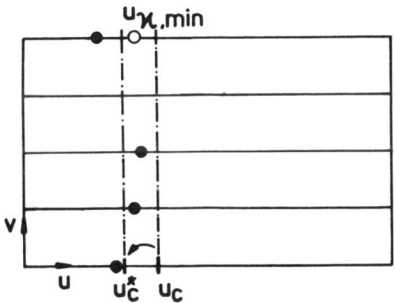

FIG. 3.8. *Cancelling of the largest parametric value of the segmentation points.*

3. The mean values of the parametric values discovered in step 1 determine the local segmentation points.

4. The smallest parametric value $u_{\kappa_{\min}}$ of the second minimum of curvature (which is at the left-hand side of the local segmentation points out of step 2) is determined.

5. The parametric value u_c of the center of gravity of all local segmentation points leads to the first boundary curve, if $u_c < u_{\kappa_{\min}}$ (see Fig. 3.7).

6. If $u_c > u_{\kappa_{\min}}$ the center of gravity u_c is moved to u_c^* by successive cancelling of the local segmentation points with the largest parameter values

during the calculation of the center of gravity of the curvature as long as the condition in step 4 holds (see Fig. 3.8).

7. If u_c^* is determined, go to step 1 and continue with $u = u_c^*$.

With this procedure we obtain that each new patch has no more than one minimum of curvature on each discretized parametric line and we get a minimal number of patches with respect to the geometric properties postulated above. Afterwards the given surface is segmented in v direction analogously.

The strategy for bicubic surfaces can be easily extended to quintic surfaces. Because of the larger numbers of the (allowed) minima of curvature, steps 1–3 must be changed as follows:

1. To determine the first four minima of curvature at each discretized parametric line $u = $ const.

2. To determine the local segmentation points as mean values of the first four parametric values discovered in step 1.

3. To determine the smallest parameter value $u_{\kappa_{min}}$ of the fourth (furthest at the right-hand side) minimum of curvature on each parametric line.

The splitting of one uniform rational B-spline surface of degree 5 with (3×4) rational B-spline segments by help of the developed procedure is demonstrated in Figs. 3.9a,b. We get (6×2) new rational Bézier or rational B-spline patches. If two parameter values of the new boundary curves enclose more than one knot of the given knot vector, the new patch is a rational B-spline patch; otherwise it is a rational Bézier patch (see Fig. 3.2). The given surface is drawn in Fig. 3.9a, while Fig. 3.9b contains the surface with its new natural, geometrically oriented boundary curves.

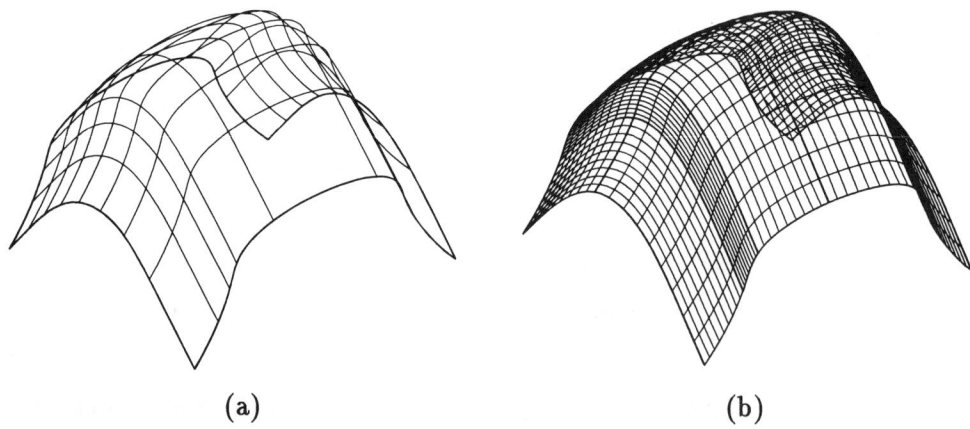

(a)   (b)

FIG. 3.9. *Splitting of a B-spline surface into Bézier or B-spline patches.* (a) *Given rational B-spline surface with nonuniform knots.* (b) *Splitting of surface from* (a) *to Bézier and B-spline patches.*

3.3.3. Approximation Strategy for Degree Reduction.

The given surface $\mathbf{X}$ may have the polynomial degree (n, m) while the required surface

Y will have the polynomial degree $(3,3)$ or $(5,5)$. The approximation process will be subdivided into two steps:

Step I: The approximation of the boundary curves of the given surface **X**.

Step II: The approximation of the interior of the surface **X**.

First we assume that the two surfaces **X** and **Y** may have the same corner points

$$(3.14) \qquad\qquad \mathbf{X}(i,k) = \mathbf{Y}(i,k) \qquad (i,k = 0,1).$$

Then we can start with Step I.

Step I. For the approximation of the boundary curves, we assume that the corresponding curves fulfill conditions of contact of order 1 in the cubic case (or conditions of contact of order 2 in the quintic case; see (3.1) or (3.4a–c)). To describe this step, for example, we pick out the boundary curve $\mathbf{X}(u,0)$. Analogously to §3.2.3 we choose on this curve $r+1$ equidistant points $\mathbf{P}_i$ with parameter values u_i $(r > n)$, and get the error vectors $\mathbf{d}_i = \mathbf{P}_i - \mathbf{Y}(u_i,0)$. The goal is to minimize the total error sum

$$(3.15) \qquad\qquad d = \sum_{i=0}(\mathbf{d}_i)^2.$$

The error vectors $\mathbf{d}_i$ between corresponding points on **X** and on **Y** are in general not perpendicular to the approximating boundary curve; thus the total error sum d is too large. We can reduce the total error (3.15) by the parameter correction (3.9a,b) (see Fig. 3.4).

Step II. Now we approximate the interior of the given surface $\mathbf{X}(u,v)$ while the approximating boundary curves remain unchanged. We assume that the given surface **X** and the approximation surface **Y** fulfill the conditions of contact of order 1 in the bicubic case (or conditions of contact of order 2 in the biquintic case; see (3.10), (3.11)) in the corner points.

The basis of the approximation process is the choice of points $\mathbf{P}_i$ on the given surface. To transfer enough information from the given surface to the approximation surface, it is necessary that these points are (nearly) equidistantly spaced on the given surface patch.

First we determine equidistant points on each boundary curve; then the corresponding parameter values will be joined by lines (see Fig. 3.10). The parameter values of the intersection points of these lines determine the points $P(u_i, v_j)$ on the given surface which are in general nearly equidistant and can be used to start our approximation process.

In general this method leads to a well-distributed set of points. Nevertheless, for large differences between the weights, the introduced method can also lead to a set of *not* well-distributed points on the given surface. For this case, other grid generations are in development.

3.3.4. Reduction of a Bicubic Surface. The goal of this procedure is to approximately convert the given surface described by (3.12) to the Bézier surfaces described by (3.13). After the splitting procedure, the given surface

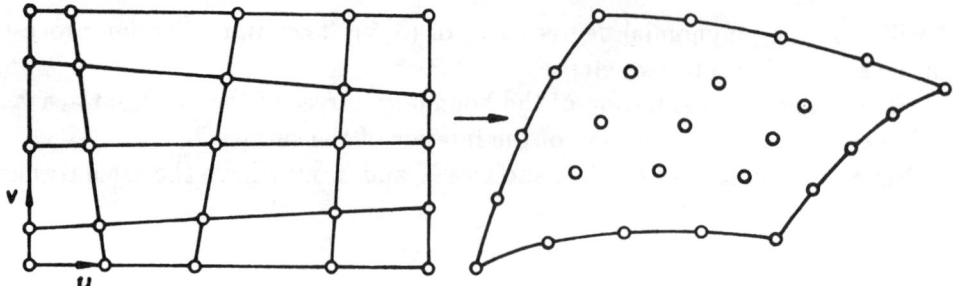

FIG. 3.10. *Choice of the parameter values on the given surface.*

patches have a Bézier or a B-spline representation. For a Bézier representation, we assume

$$(3.16) \quad \mathbf{X}(u,v) = \frac{\sum_{i=0}^{n}\sum_{k=0}^{m}\beta_{ik}\mathbf{V}_{ik}B_i^n(u)B_k^m(v)}{\sum_{i=0}^{n}\sum_{k=0}^{m}\beta_{ik}B_i^n(u)B_k^m(v)}, \qquad (u,v) \in [0,1],$$

with Bernstein polynomials $B_i^l(t)$ of degree $l = n$, respectively, $l = m$, $\mathbf{V}_{ik}$ as given Bézier points, and the weights β_{ik}. Analogously a B-spline segment may have the representation

$$(3.17) \qquad \mathbf{X}(u,v) = \frac{\sum_{i=0}^{s}\sum_{k=0}^{\sigma}\beta_{ik}\mathbf{V}_{ik}N_{iN}(u)N_{kM}(v)}{\sum_{i=0}^{s}\sum_{k=0}^{\sigma}\beta_{ik}N_{iN}(u)N_{kM}(v)}$$

with $u \in [u_0, u_1]$, $v \in [v_0, v_1]$, and the B-spline basis function $N_{il}(t)$ of order $l = N$, respectively, $l = M$. If we compare (3.16) with (3.17) we have $n = N-1$, $m = M-1$ as polynomial degrees.

Such a surface patch should be converted into the bicubic Bézier patch

$$(3.18) \qquad \mathbf{Y}(u,v) = \sum_{i=0}^{3}\sum_{k=0}^{3} \mathbf{W}_{ik}B_i^3(u)B_k^3(v).$$

The boundary points $\mathbf{W}_{00}$, $\mathbf{W}_{03}$, $\mathbf{W}_{30}$, $\mathbf{W}_{33}$ coincide with the corner points of the given patch. For the conversion of the boundary curves, the new Bézier points are determined by contact of order 1 conditions. For instance, for curve 1 (see Fig. 3.11), we get from (3.4a–c) with $u = 0$, $v \in [0,1]$,

$$(3.19) \quad \begin{aligned} &\mathbf{W}_{00} = \mathbf{V}_{00}, \quad \mathbf{W}_{03} = \mathbf{V}_{0m}, \quad \mathbf{W}_{01} = \mathbf{V}_{00} + \lambda_1(\mathbf{V}_{01} - \mathbf{V}_{00}), \\ &\mathbf{W}_{02} = \mathbf{V}_{0m} + \lambda_2(\mathbf{V}_{0,m-1} - \mathbf{V}_{0m}). \end{aligned}$$

The parameters λ_i and parameter correction (3.9a,b) are used for minimization of the total error sum of the boundary curve with help of least square methods [15]. After Step I the Bézier points $\mathbf{W}_{0k}$, $\mathbf{W}_{i0}$, $\mathbf{W}_{3k}$, $\mathbf{W}_{i3}$ $(i,k = 1,2)$ are determined by the scalar parameters λ_j (see Fig. 3.11). Now the boundary curves of the given and of the approximating surface have common tangents in the corner points.

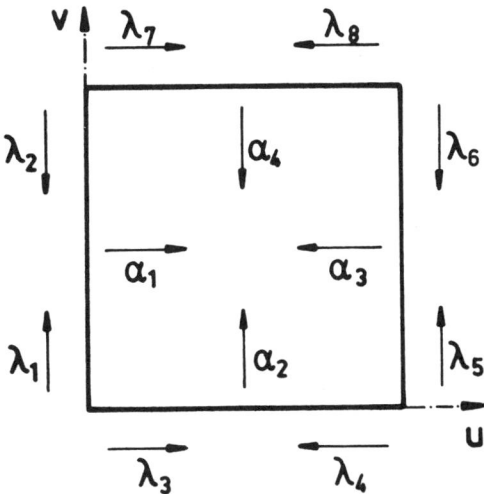

FIG. 3.11. *Effect of the introduced approximation parameters in the bicubic case.*

For evaluation of the unknown inner Bézier points $\mathbf{W}_{11}$, $\mathbf{W}_{12}$, $\mathbf{W}_{21}$, $\mathbf{W}_{22}$, we further assume that the following conditions will hold for the cross derivatives at the boundary curves:

$$
\begin{aligned}
\text{edge 1:} \quad & \mathbf{Y}_u(0,v) = \alpha_1(v)\mathbf{X}_u(0,v), \\
\text{edge 2:} \quad & \mathbf{Y}_v(u,0) = \alpha_2(u)\mathbf{X}_v(u,0), \\
\text{edge 3:} \quad & \mathbf{Y}_u(1,v) = \alpha_3(v)\mathbf{X}_u(1,v), \\
\text{edge 4:} \quad & \mathbf{Y}_v(u,1) = \alpha_4(u)\mathbf{X}_v(u,1),
\end{aligned}
$$
(3.20)

where the unknown functions α_i are determined at the boundary curves by the (known) parameters λ_k, corresponding to (3.19). As described in Fig. 3.11, the following boundary conditions hold for the α_i:

(3.21)

$$
\begin{aligned}
\text{edge 1:} \quad (u=0): \quad & \alpha_1(v=0) = \frac{3}{n}\lambda_3, \quad & \alpha_1(v=1) = \frac{3}{n}\lambda_7, \\
\text{edge 2:} \quad (v=0): \quad & \alpha_2(u=0) = \frac{3}{m}\lambda_1, \quad & \alpha_2(u=1) = \frac{3}{m}\lambda_5, \\
\text{edge 3:} \quad (u=1): \quad & \alpha_3(v=0) = \frac{3}{n}\lambda_4, \quad & \alpha_3(v=1) = \frac{3}{n}\lambda_8, \\
\text{edge 4:} \quad (v=1): \quad & \alpha_4(u=0) = \frac{3}{m}\lambda_2, \quad & \alpha_4(u=1) = \frac{3}{m}\lambda_6.
\end{aligned}
$$

We set up the functions α_i with respect to (3.21) as quadratic functions

$$\alpha_1(v) = \frac{3}{n}\lambda_3 B_0^2(v) + \omega_1 B_1^2(v) + \frac{3}{n}\lambda_7 B_2^2(v),$$

$$\alpha_2(u) = \frac{3}{m}\lambda_1 B_0^2(u) + \omega_2 B_1^2(u) + \frac{3}{m}\lambda_5 B_2^2(u),$$

(3.22)

$$\alpha_3(v) = \frac{3}{n}\lambda_4 B_0^2(v) + \omega_3 B_1^2(v) + \frac{3}{n}\lambda_8 B_2^2(v),$$

$$\alpha_4(u) = \frac{3}{m}\lambda_2 B_0^2(u) + \omega_4 B_1^2(u) + \frac{3}{m}\lambda_6 B_2^2(u),$$

with ω_i as parameters which will be used for optimization. If we insert (3.22) into the continuity conditions (3.20), we obtain the following vector-valued linear system for the unknown Bézier points $\mathbf{W}_{11}$, $\mathbf{W}_{12}$, $\mathbf{W}_{21}$, $\mathbf{W}_{22}$:

(3.23)
$$\begin{pmatrix} B_1^3(v) & B_2^3(v) & 0 & 0 \\ B_1^3(u) & 0 & B_2^3(u) & 0 \\ 0 & 0 & B_1^3(v) & B_2^3(v) \\ 0 & B_1^3(u) & 0 & B_2^3(u) \end{pmatrix} \begin{pmatrix} \mathbf{W}_{11} \\ \mathbf{W}_{12} \\ \mathbf{W}_{21} \\ \mathbf{W}_{22} \end{pmatrix} = \begin{pmatrix} \frac{n}{3}\omega_1\mathbf{N}_1(v) + \mathbf{Q}_1(v) \\ \frac{m}{3}\omega_2\mathbf{N}_2(u) + \mathbf{Q}_2(u) \\ \frac{n}{3}\omega_3\mathbf{N}_3(v) + \mathbf{Q}_3(v) \\ \frac{m}{3}\omega_4\mathbf{N}_4(u) + \mathbf{Q}_4(u) \end{pmatrix},$$

where the $\mathbf{N}_i$, $\mathbf{Q}_i$ are expressions with well-known quantities. The linear system has rank 3; thus (3.23) describes the Bézier points $\mathbf{W}_{12}$, $\mathbf{W}_{21}$, $\mathbf{W}_{22}$ for a (freely chosen) point (u_0, v_0), for example, as functions of the parameters $\omega_1, \cdots, \omega_4$ and the Bézier point $\mathbf{W}_{11}$. One of the vanishing determinants of system (3.23) leads to a (vector-valued) condition determining $\omega_i = \omega_i(\omega_1)$ ($i = 2, 3, 4$). Thus the whole problem is reduced to four unknown linear variables: ω_1 and the components of $\mathbf{W}_{11}$.

The total error sum d depends on these four variables and that the following conditions hold for the minimum of the total error sum:

(3.24)
$$\frac{\partial d}{\partial \mathbf{W}_{11}} = 0, \qquad \frac{\partial d}{\partial \omega_1} = 0.$$

These conditions provide a linear system for ω_1 and the components of $\mathbf{W}_{11}$. Additionally, the parameter transformation [17], [20], [27], analogously to (3.9b), is used to reduce the approximation error. Figure 3.12 describes the procedure for the approximation of the interior surface. For the boundary curve, see Fig. 3.4.

After the approximation process, the given surface $\mathbf{X}$ and the approximation surface $\mathbf{Y}$ have the following geometric correspondence. They have common tangent planes in each corner point $\mathbf{V}_{00} = \mathbf{W}_{00}$, $\mathbf{V}_{0m} = \mathbf{W}_{03}$, $\mathbf{V}_{n0} = \mathbf{W}_{30}$, $\mathbf{V}_{nm} = \mathbf{W}_{33}$, and with respect to (3.20) parallel tangents in the corresponding boundary points $(0, v_0)$, $(u_0, 0)$, $(1, v_0)$, $(u_0, 1)$. Nevertheless, in general the converted set of patches is only C^0 continuous. Therefore, additional algorithms are used to reduce the jumps of the normal vectors along the boundary curves to less than two degrees.

In general, the segmentation of the given surface with the help of the geometric boundary curves (as described in §3.2) is not sufficient to fulfill a given maximal error condition. Therefore, additional segmentation strategies are necessary. First we determine the point $\mathbf{P_{max}} = \mathbf{P}(u_i, v_j)$ with the largest approximation error. As approximation errors the following criteria can be used:

(i) The maximal distance of $\mathbf{P_{max}}$ from the approximation surface;

(ii) The maximal deviation of the normal vectors in $\mathbf{P_{max}}$ and the corresponding point on the approximation surface.

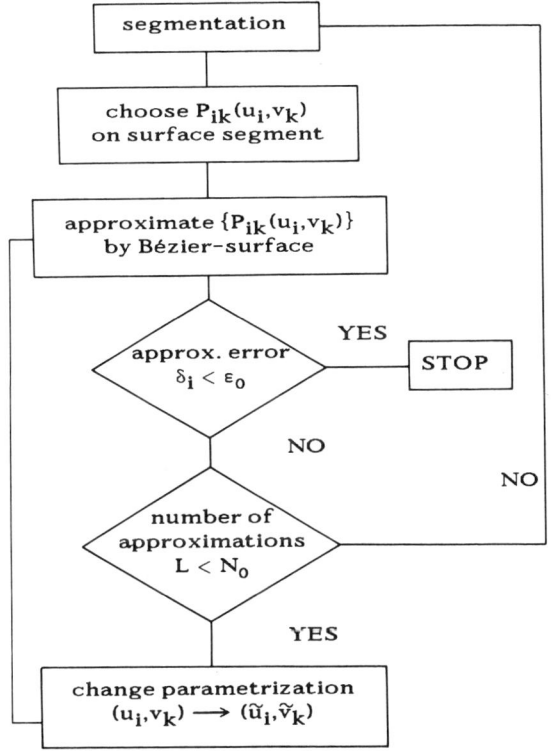

FIG. 3.12. *Procedure for approximation of the interior of a surface with parameter correction.*

To determine the direction of the segmentation line, we consider the star-shaped neighboring points to $\mathbf{P_{max}}$: $\mathbf{P}(u_{i-1}, v_j)$, $\mathbf{P}(u_{i+1}, v_j)$, $\mathbf{P}(u_i, v_{j-1})$, $\mathbf{P}(u_i, v_{j+1})$, and the corresponding points on the approximation surface. If the largest error values appear at the points $\mathbf{P}(u_{i-1}, v_j)$, $\mathbf{P}(u_{i+1}, v_j)$, the given surface is split in v_j directions. Thus the points with the largest error values are on the new boundary curves.

Figures 3.13 and 3.14 show two examples of degree reduction. In Fig. 3.13 an $(8,8)$ Bézier surface is converted to four segments of bicubic Bézier patches.

Both surfaces are drawn one above the other. The parametric lines of the given surface are drawn as solid lines, while the parametric lines of the approximation patches are drawn broken. The differences of corresponding parametric lines demonstrate the effect of the reparameterization. In Fig. 3.14 a rational B-spline surface of degree $(5, 5)$ with 3×4 segments is converted to 6×2 bicubic Bézier patches. The maximum error at the boundary curve is less than 0.09 percent; in the interior the maximum error is less than 0.12 percent. The normal vectors at the boundary curves differ with the maximum angle of 1.86 degrees.

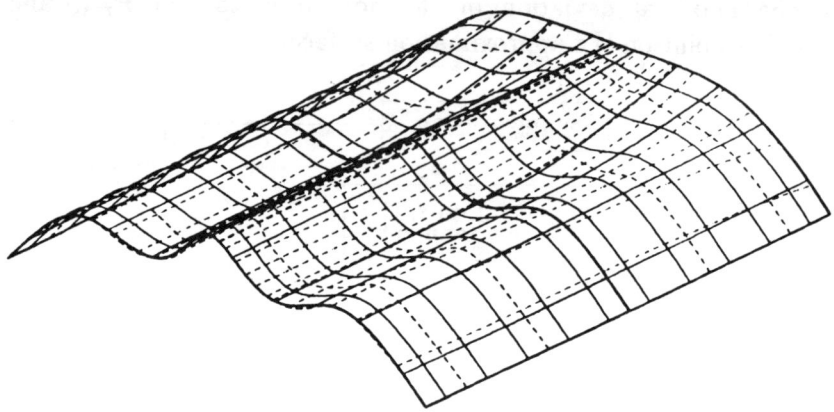

FIG. 3.13. *Conversion of an (8,8) Bézier surface to four bicubic patches.*

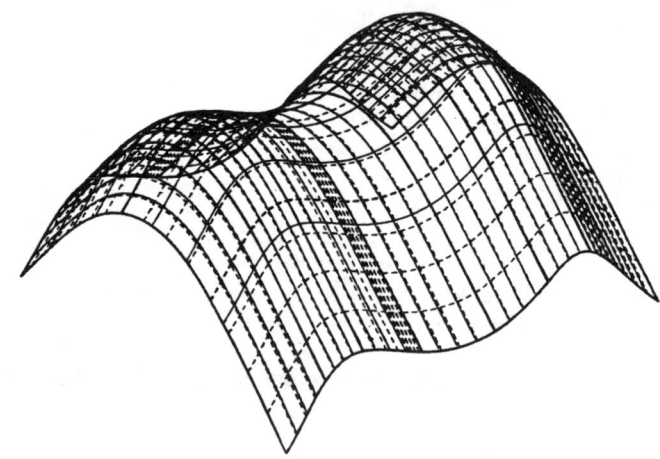

FIG. 3.14. *Conversion of a biquintic rational B-spline surface to 6×2 bicubic Bézier patches.*

Brode [4] has worked out a comparison between the methods of Dannenberg and Nowacki [7], Bardis and Patrikalakis [3], and our method. Because of reparameterization our method in general leads to the smallest number of new patches. As an example, in [4] the $(9, 9)$ integral Bézier surface (Fig. 3.15) was converted to bicubic patches (see Table 3.1).

TABLE 3.1

Conversion of the (9 × 9) *Bézier surface in Fig.* 3.15 *to a set of bicubic Bézier patches (as bases [4] and calculations with the original implementation).*

Method	max. error	number of patches	continuity
BP 89	0.01	144	C^2
DN 85	0.01	25	C^0
HSW 89	0.05	8	C^0
	0.01	12	C^0
LA 88	0.01	24	C^0
	0.1	56	C^0
	0.01	40	C^1
	0.1	128	C^1

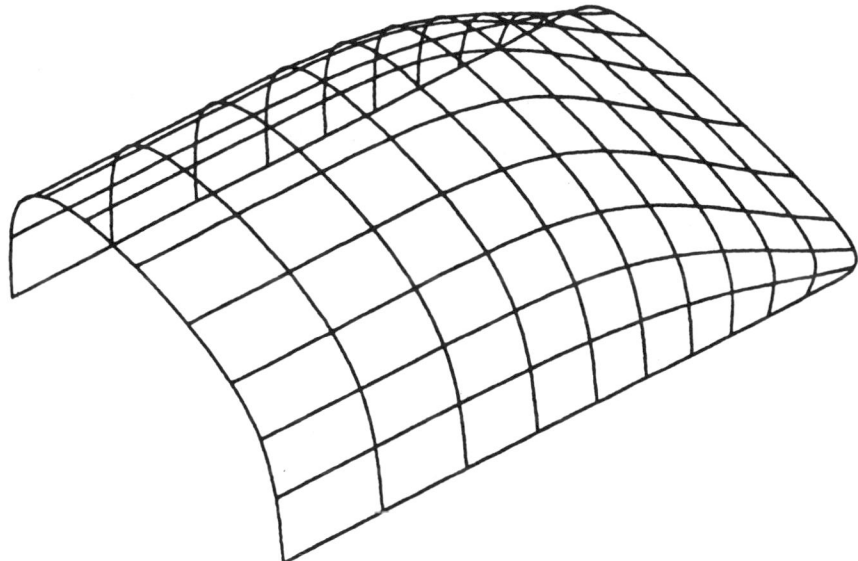

FIG. 3.15. *The* (9,9) *Bézier surface used in [4] for comparing the different methods.*

3.3.5. Reduction to a Biquintic Surface.   Now the required surface may have the representation

$$(3.25) \qquad \mathbf{Y}(\mu,\nu) = \sum_{i=0}^{5}\sum_{k=0}^{5} \mathbf{W}_{ik} B_i^5(\mu) B_k^5(\nu).$$

Again the approximating boundary curves are evaluated first with methods developed in §3.2.

After this first step of the approximation process, the Bézier points $\mathbf{W}_{0k}$, $\mathbf{W}_{i0}$, $\mathbf{W}_{5k}$, $\mathbf{W}_{i5}$ $(i, k = 1, 2, 3, 4)$ are determined by the scalar parameters λ_j, μ_j (see Fig. 3.16). In the vertices the boundary curves of the given and the approximating surface have common tangents and the same curvature.

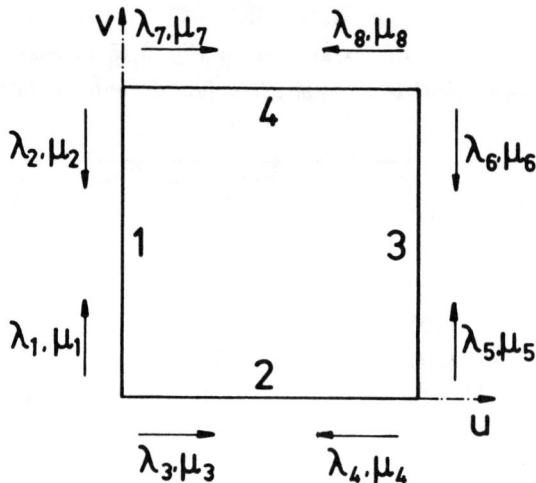

FIG. 3.16. *Effect of the introduced approximation parameters in the biquintic case.*

Now we choose a special case of (3.11) as continuity conditions at the boundary curves of the given and the approximating surface. For example, at the boundary curves 1, 2 the following cross conditions will hold:

curve 1:

$$\mathbf{Y}_u(0, v) = {}^1\nu_{10}(v)\mathbf{X}_u(0, v),$$

(3.26) $$\mathbf{Y}_{uu}(0, v) = ({}^1\nu_{10}(v))^2\mathbf{X}_{uu}(0, v) + {}^1\nu_{20}(v)\mathbf{X}_u(0, v),$$

$$\mathbf{Y}_{uv}(0, v) = {}^1\bar{\nu}(v)\mathbf{X}_{uv}(0, v) + {}^1\bar{\sigma}(v)\mathbf{X}_v(0, v);$$

curve 2:

$$\mathbf{Y}_v(u, 0) = {}^2\sigma_{01}(u)\mathbf{X}_v(u, 0),$$

$$\mathbf{Y}_{vv}(u, 0) = ({}^2\sigma_{01}(u))^2\mathbf{X}_{vv}(u, 0) + {}^2\sigma_{02}(u)\mathbf{X}_v(u, 0),$$

$$\mathbf{Y}_{uv}(u, 0) = {}^2\bar{\nu}(u)\mathbf{X}_{uv}(u, 0) + {}^2\bar{\sigma}(u)\mathbf{X}_u(u, 0).$$

The unknown functions ${}^k\nu_{ij}$, ${}^k\sigma_{ij}$, ${}^k\bar{\nu}$, ${}^k\bar{\sigma}$, ($k = 1, \cdots 4$; $i, j = 0, \cdots 2$) can be set up as linear or as quadratic functions. These functions must fulfill the boundary values of μ_k, ν_k which are evaluated for the boundary curves. Therefore, we can introduce the following functions.

curve 1:

$$ {}^1\nu_{10}(v) = \frac{5}{n}\lambda_3 B_0^2(v) + {}^1\omega_8 B_1^2(v) + \frac{5}{n}\lambda_7 B_2^2(v),$$

$$ {}^1\nu_{20}(v) = \frac{20}{n}\left(\mu_3 + (\lambda_3)^2\frac{5(n-1)}{4n} - 2\lambda_3\right)B_0^2(v) + {}^1\omega_7 B_1^2(v)$$

(3.27) $$\qquad + \frac{20}{n}\left(\mu_7 + (\lambda_7)^2\frac{5(n-1)}{4n} - 2\lambda_7\right)B_2^2(v),$$

$$ {}^1\bar{\nu}(v) = \frac{25}{nm}\left(\lambda_1\lambda_3 B_0^1(v) + \lambda_2\lambda_7 B_1^1(v)\right),$$

$$ {}^1\sigma(v) = {}^1\omega_9 B_2^2(v)$$

with the Bernstein polynomials $B_k^i(v)$ and the unknowns ${}^1\omega_7$, ${}^1\omega_8$, ${}^1\omega_9$. λ_3, λ_7, μ_3, μ_7 are determined from the approximation of the boundary curves. For ${}^2\sigma_{01}(u)$, ${}^2\sigma_{02}(u)$, ${}^2\bar{v}(u)$, ${}^2\bar{\sigma}(u)$, analogous functions can be set up.

Additionally we assume that the twist vectors $\mathbf{X}_{uv}$, $\mathbf{Y}_{uv}$ of the given and the required surface are parallel in the vertices. The Bézier points $\mathbf{W}_{11}$, $\mathbf{W}_{14}$, $\mathbf{W}_{41}$, $\mathbf{W}_{44}$ are determined out of this parallelism of the twist vectors by conditions like

$$\mathbf{W}_{11} = \lambda_1\lambda_3(\mathbf{V}_{00} - \mathbf{V}_{01} - \mathbf{V}_{10} + \mathbf{V}_{11}) + \mathbf{W}_{01} + \mathbf{W}_{10} - \mathbf{W}_{00}.$$

Now we have to record the still undetermined 12 interior Bézier points. From (3.26) and analogous conditions at the other boundary curves, we obtain a vector-valued linear system for these 12 unknown Bézier points. A linear system for the Bézier points $\mathbf{W}_{22}$, $\mathbf{W}_{23}$, $\mathbf{W}_{32}$, $\mathbf{W}_{33}$ can be isolated out of this system. This system has a singular coefficient matrix analogous to (3.23). One of these vanishing determinants of the system leads to three scalar constraint conditions. If we now choose (u_0, v_0) out of this smaller system, the Bézier points $\mathbf{W}_{22}$, $\mathbf{W}_{23}$, $\mathbf{W}_{32}$ can be recorded as functions of the parameters ${}^i\omega_k$ and the components of $\mathbf{W}_{33}$. The points $\mathbf{W}_{12}$, $\mathbf{W}_{13}$, $\mathbf{W}_{21}$, $\mathbf{W}_{31}$, $\mathbf{W}_{42}$, $\mathbf{W}_{43}$, $\mathbf{W}_{24}$, $\mathbf{W}_{34}$ can be evaluated from the big system for the unknown 12 Bézier points. Thus we can determine the total error sum as a function of these unknowns. The components of $\mathbf{W}_{33}$ and the parameters ${}^1\omega_7$, ${}^1\omega_9$, ${}^2\omega_9$, ${}^3\omega_9$, ${}^4\omega_9$ are linear unknowns, while the parameters ${}^1\omega_8$, ${}^2\omega_8$, ${}^3\omega_8$, ${}^4\omega_8$ are nonlinear unknowns. To minimize the total error sum, nonlinear approximation methods are used combined with suitable parameter corrections as described in §3.2. At the end of this approximation process, the given surface $\mathbf{X}$ and the approximation surface $\mathbf{Y}$ have the following geometric correspondence. They have common tangent planes and the same Dupinian indicatrix in each corner point

$$\mathbf{V}_{00} = \mathbf{W}_{00}, \quad \mathbf{V}_{0m} = \mathbf{W}_{05}, \quad \mathbf{V}_{n0} = \mathbf{W}_{50}, \quad \mathbf{V}_{nm} = \mathbf{W}_{55}.$$

Figure 3.17a shows an example of degree reduction in the biquintic case. A $(7,7)$ Bézier surface is converted into one $(5,5)$ Bézier patch. The example leads to the following error values: for the boundary curves $\alpha_1 = 0.08$ percent, for the interior of the surface $\epsilon_1 = 0.31$ percent. The directions of the normal vectors differ with a maximal value of $\gamma_1 = 0.88$ degrees.

The quality of the approximate conversion can be checked by curvature plots. Figure 3.17b contains a plot of the curvature of a parametric line of the surface out of Fig. 3.17a. The curvature diagrams of the given surfaces are drawn by solid lines; the diagrams of the approximation surfaces are drawn by broken lines.

After the conversion of each patch the converted set of patches is in general only C^0 continuous. In the corner points of the patches we have G^2 continuity. Additionally with the help of the assumption on the cross derivatives, we get G^2 continuity for one (arbitrarily chosen) parametric line crossing the boundary curves. Therefore, further algorithms must be used to reduce the jumps of the

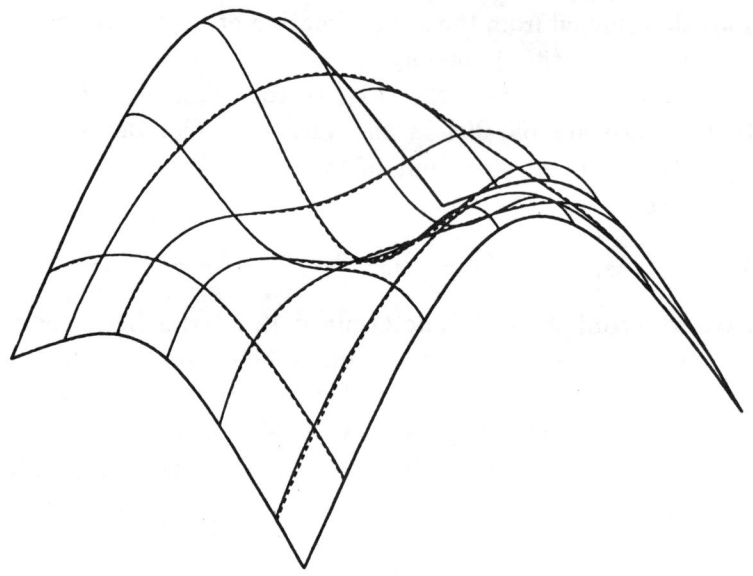

FIG. 3.17a. *Conversion of a (7,7) Bézier surface to a (5,5) Bézier surface.*

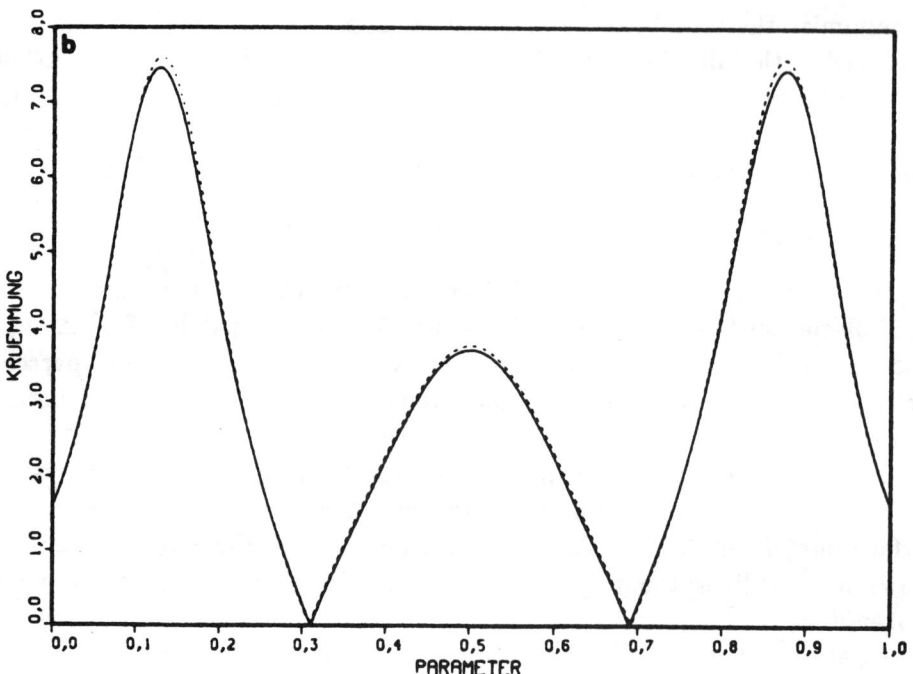

FIG. 3.17b. *Curvature of the line $u = 0.5$ of the conversion $(7,7) \rightarrow (5,5)$.*

normal vectors along the boundary curves to less than two degrees. In [27] the use of the higher mixed derivatives is proposed to get continuous derivatives along the boundary curves.

Wassum [27], [28] has additionally introduced higher cross derivatives to reduce the number of unknowns. Approximate conversion algorithms to higher degree can be developed with this method (for instance to $(7,7)$ or $(9,9)$). Then higher contact of order is obtained in the corners of the patches, for instance, biquintic surfaces are G^1 continuous, the biseptic surfaces G^2 continuous, and so on.

3.3.6. Merging of Surface Patches and Reduction to Higher Degree.
The method of degree reduction can also be used for merging a set of Bézier or B-spline patches. If there are small given patches (which followed from the construction process), this lot of patches can be merged with the help of our segmentation strategy to a smaller number of Bézier patches (see Figs. 3.18a–c). In this case the polynomial degree can be changed, but it can also be equal to the given surfaces.

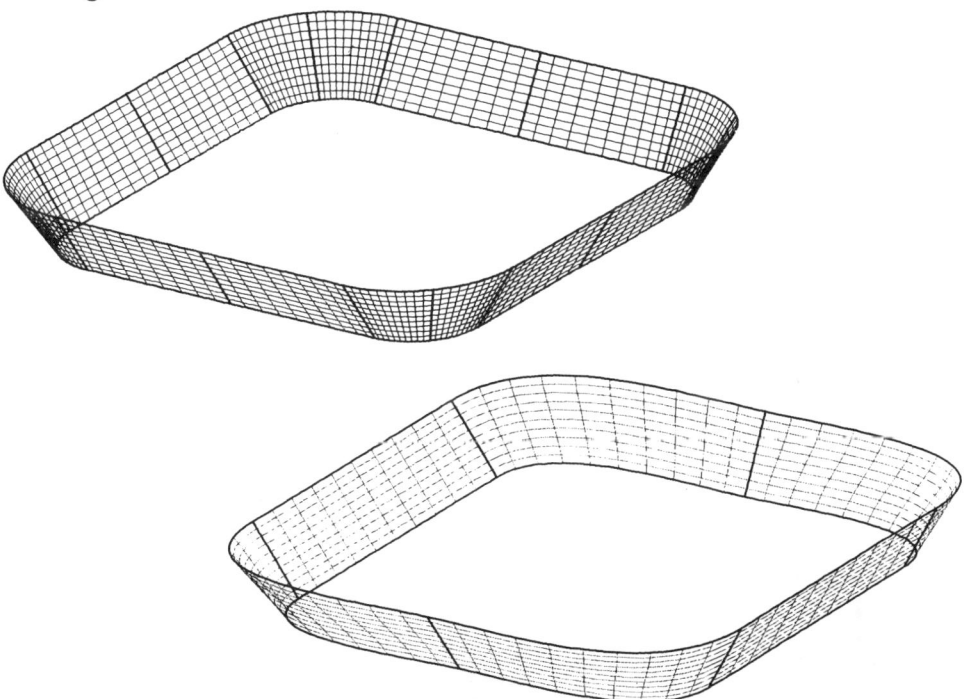

FIG. 3.18a. *Merging* 16 *patches of degree* $(5,1)$ *(top) to 6 bicubic patches (bottom)* (*maximum is* 0.18 *mm; total extension is* 40 *cm*).

3.3.7. Conversion of Triangular Patches.
Figures 3.18a–c contain a triangular patch which is converted to a part of a rectangular patch. Certainly our method can also be extended to conversion of triangular patches to triangular patches. If the triangular patches are given as degenerated

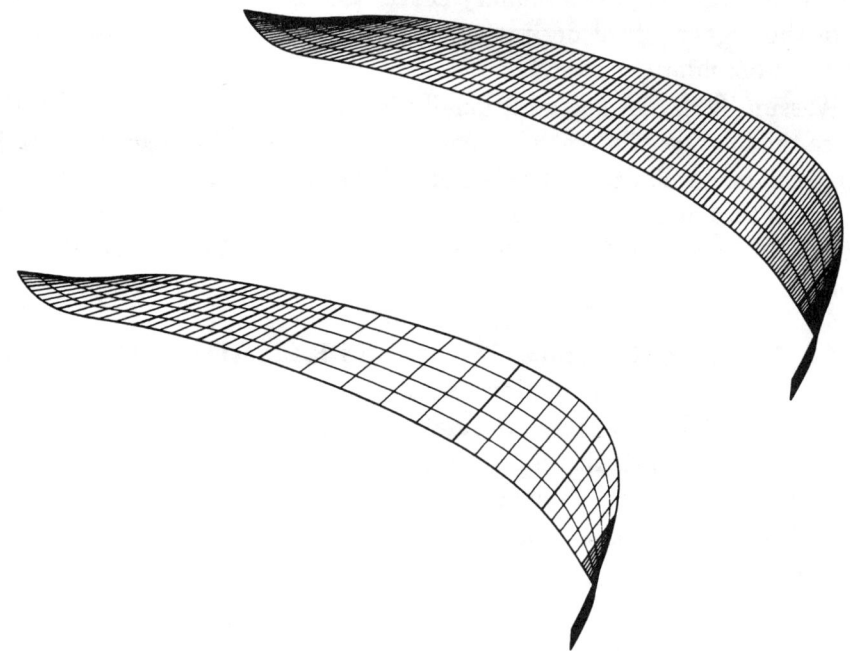

FIG. 3.18b. *Merging 29 bicubic patches to 11 bicubic patches (maximum error is 0.1 mm; total extension is 25 cm).*

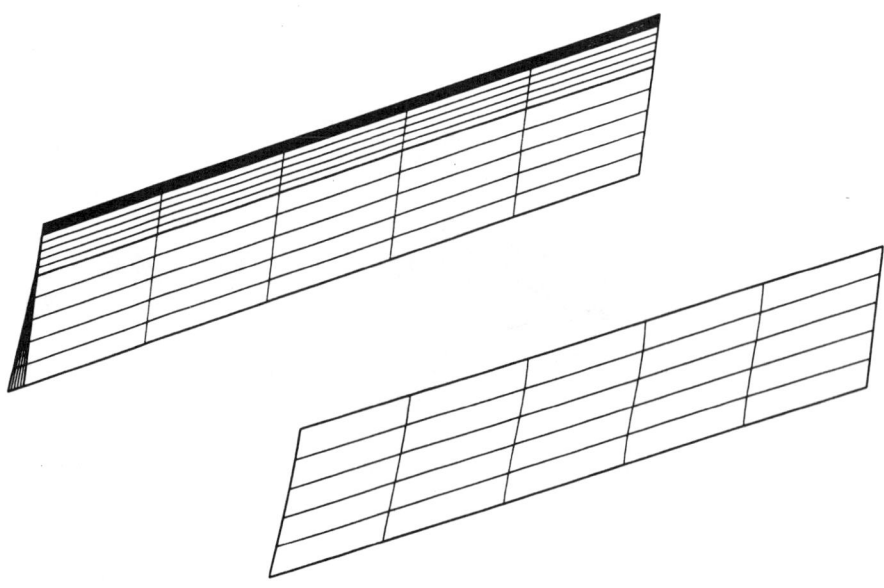

FIG. 3.18c. *Merging of six patches of degree (5,1) to one bicubic patch. At the left-hand side small triangular patches are given. (Maximum error is 0.2 mm in interior, 0.47 mm at boundary; maximum extension is 40 cm.)*

rectangular patches (see [12]), where the length of one boundary curve of a rectangular patch tends to zero or the angle between two boundary curves of a rectangular patch tends to 180 degrees, our method can be used directly. Only in the degenerated case first mentioned do we have to introduce suitable assumptions for the cross derivatives.

If the triangular Bézier technique is used for patch description, a new problem arises with respect to the rectangular case. The cross derivatives of the boundary curves are not independent; therefore, boundary conditions must be fulfilled. Thus in the cubic case, only the one inner Bézier point is available for approximation purposes, while in the quintic case the three "inner" Bézier points $\mathbf{b}_{122}$, $\mathbf{b}_{212}$, $\mathbf{b}_{221}$ and three scalar parameters (from the first derivatives) are available for optimization. The Bézier points $\mathbf{b}_{113}$, $\mathbf{b}_{311}$, $\mathbf{b}_{131}$ are determined by parallelity to the twist vectors in the vertices. Figure 3.19 contains a triangular degree reduction from degree 7 to degree 5. The given and the approximating surface are drawn on top of one another and marked by solid lines and dashed lines, respectively.

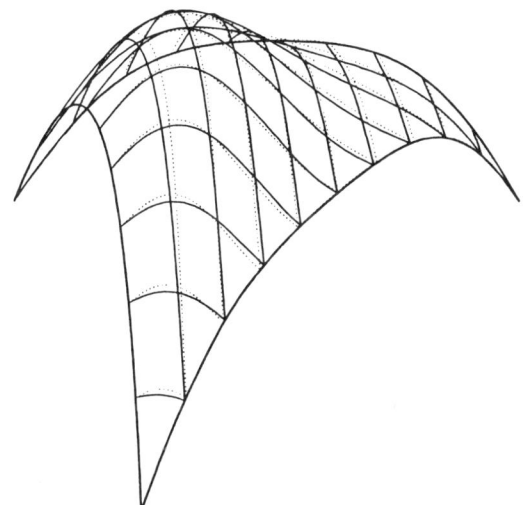

FIG. 3.19. *Conversion of a triangular Bézier patch of degree 7 to a quintic one.*

3.3.8. Approximation Strategy for Trimmed Surfaces. We assume that the curves $\mathbf{C}_i$ on the given trimmed surface are described in the parametric domain by nonuniform rational B-spline representations

$$(3.28a) \qquad \mathbf{C}_i(t) = \begin{pmatrix} u_i(t) \\ v_i(t) \end{pmatrix} = \frac{\sum_{j=0}^{p_i} \beta_{ij} \mathbf{d}_{ij} N_{jk_i}(t)}{\sum_{j=0}^{p_i} \beta_{ij} N_{jk_i}(t)}$$

with k_i as order of the curves $\mathbf{C}_i$. The knot vectors of $\mathbf{C}_i$ may be denoted by $\mathbf{T}$ with

$$(3.28b) \qquad \mathbf{T}_i = \left(t_0^i = t_1^i = \cdots t_{k-1}^i, t_k^i \cdots t_{p_i}^i, t_{p_i+1}^i = \cdots t_{p_i+k}^i \right).$$

On the converted set of (integral) Bézier surfaces, these curves may have the representations

$$(3.29a) \qquad \mathbf{C}_i^*(t) = \sum_{j=0}^{\bar{p}_i} \mathbf{d}_{ij}^* N_{jl_i}(t)$$

with l_i as order of the B-spline curves and the knot vectors

$$(3.29b) \qquad \bar{\mathbf{T}}_i = \left(\vec{t}_0^{\,i} = \vec{t}_1^{\,i} = \cdots \vec{t}_{l-1}^{\,i}, \vec{t}_l^{\,i} \cdots \vec{t}_{p_i}^{\,i}, \vec{t}_{p_i+1}^{\,i} = \cdots \bar{t}_{\bar{p}_i+l}^{\,i} \right).$$

The de Boor points $\mathbf{d}_{ij}^*$ are unknown as well as the knots $\vec{t}_j^{\,i}$ of the knot vector. For suitably chosen knots in the knot vector, the B-spline representation changes to a representation of Bézier splines [8], [12].

We will describe the approximation strategy for a trimmed surface; details of the approximation procedure follow below. As an example, we choose a B-spline surface F with (4×3) segments. On this surface F there might be two given spline curves C_i (see Fig. 3.20).

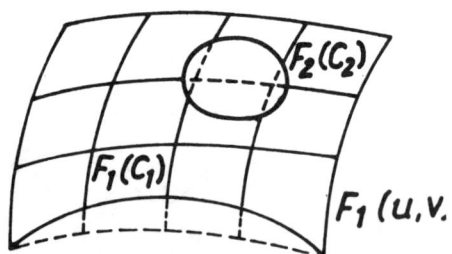

FIG. 3.20. *Example of a trimmed surface.*

First, we determine, as described in §3.3, new boundary curves. Afterwards we approximate each patch as described in §§3.3.4 and 3.3.5 by a bicubic or biquintic patch. That means we convert the given set $F_1(u, v)$ of degree (n, m) to a new set $F_2(\bar{u}, \bar{v})$ of degree $(3,3)$ (or $(5,5)$). While our approximation process uses parameter transformation, the given parameterization (u, v) is transformed in $(\bar{u}, \bar{v})$. This first step is described in Fig. 3.21a.

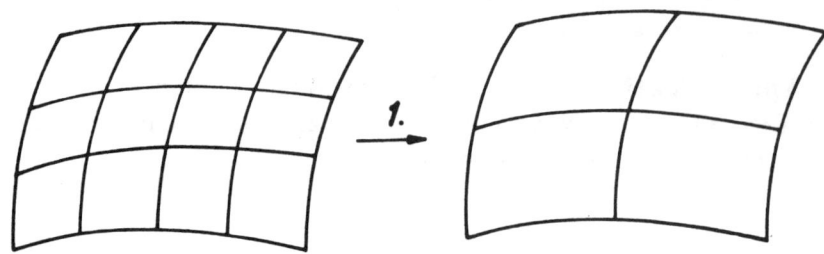

FIG. 3.21a. *Transformation of a B-spline surface F_1 to a set of bicubic Bézier surfaces F_2.*

Second, we approximate the given rational B-spline curves $C_i(t)$ in the parametric domain by integral B-spline curves $C_i^*(t)$ (see Fig. 3.21b). The absolute sum of the distances of the approximation curves $F_2(C_i^*)$ on the approximation surface F_2 from the given curves $F_1(C_i)$ is used (for details, see below) as an error norm.

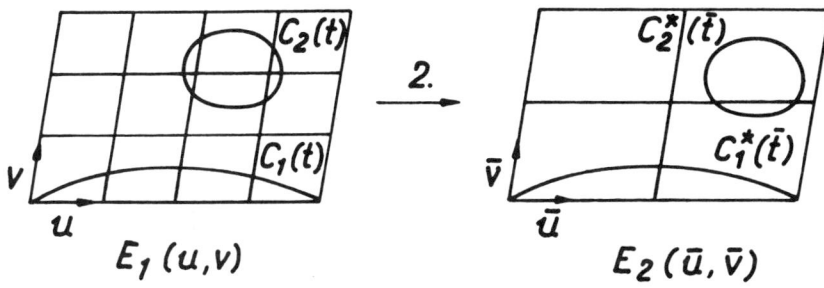

FIG. 3.21b. *Approximation of the rational B-spline curve C_i by integral B-spline curves C_i^* in the parametric domain.*

If now we additionally evaluate the intersection points of the parametric lines of the approximation surface F_2 with the curves $F_2(C_i^*)$ and cancel the undesired parts of the surface patches F_2, we get the approximation of the given trimmed surface (see Fig. 3.21c).

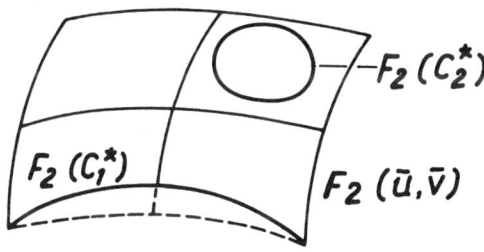

FIG. 3.21c. *Approximation of the trimmed surface.*

3.3.9. Approximate Conversion of Curves on Surfaces.

The given curves C_i may have the rational (nonuniform) B-spline representation (3.28a). i is a number of the curves with $i = 1(1)L$, with given nonuniform knot vectors as described in (3.28b). The approximation curves C_i^* (i number of curves, $i = 1(1)L$) may have the representation (3.29a) with order l_i. Unknown are the de Boor points $\mathbf{d}_{ij}^*$ and the knots of the knot vectors. The lengths of the knot vectors may be $\bar{p}_i - l_i + 2$. The approximation process runs through some steps. For instance, we will describe the approximation for one curve C.

1. First we choose in the parametric domain $E_1(u, v)$ on C_i equidistant points (u_l, v_l) $(l = 1(1)s)$. These points are mapped on the surface F_1 and we determine the points $\mathbf{P}_i$ on the curve $F_1(C_1)$ (see Fig. 3.22).

2. Now we project $\mathbf{P}_i$ onto the approximation surface F_2 with help of the surface normals of F_2. For this projection we use the parameter correction as introduced above. The foot points of the perpendiculars lead to the points

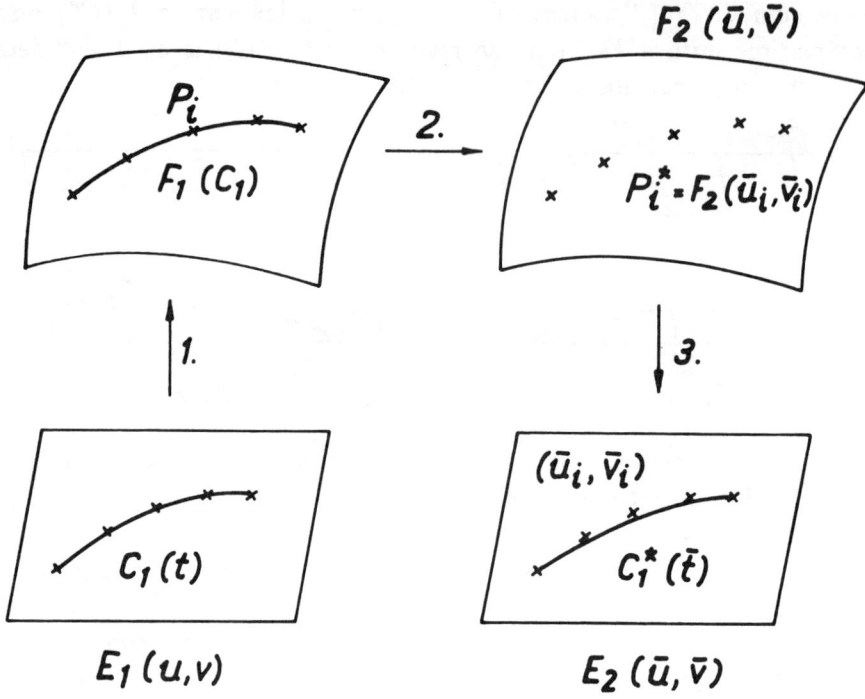

FIG. 3.22. *Schematic course of the approximation process.*

$\mathbf{P}_i^*$ on F_2 with the parameter values $(\bar{u}_l, \bar{v}_l)$, $l = 1(1)s$. They may have the (chordal) parameter values $\bar{t}_l$ with respect to the knot vector of C^*.

3. The points $(\bar{u}_l, \bar{v}_l)$ in the parametric domain $E_2(\bar{u}, \bar{v})$ shall be approximated with the B-spline curve C^* by minimizing the absolute error sum of the error vectors

$$(3.30) \qquad \delta_l = \begin{pmatrix} \bar{u}_l \\ \bar{v}_l \end{pmatrix} - C^*(\bar{t}_l^*).$$

During the approximation process the parameter values t_l are transformed to $\bar{t}_l^*$ by iterative parameter transformation as introduced in [17], [20]. At the end of the approximation procedure, the error vectors δ_l are (within a given error tolerance) perpendicular to the approximation curve C^*.

Some geometric boundary conditions are additionally required for this approximation procedure (see Fig. 3.23).

1. The starting point $\mathbf{P}_0^*$ and the end point $\mathbf{P}_s^*$ of the approximation curve $F_2(C^*)$ follow from the foot points of the perpendicular of the starting point $\mathbf{P}_0$ and the end point $\mathbf{P}_s$ of the given curve $F_1(C)$ on the approximation surface F_2.

2. The direction of the starting tangent $\mathbf{T}_0^*$ and the end tangent $\mathbf{T}_s^*$ of the approximation curve $F_2(C^*)$ is determined by (orthogonal) projection of the starting tangent $\mathbf{T}_0$ and the end tangent $\mathbf{T}_s$ of the given curve $F_1(C)$ into the tangent plane $DF_2(\mathbf{P}_0^*)$, $DF_2(\mathbf{P}_s^*)$, respectively (see Fig. 3.23).

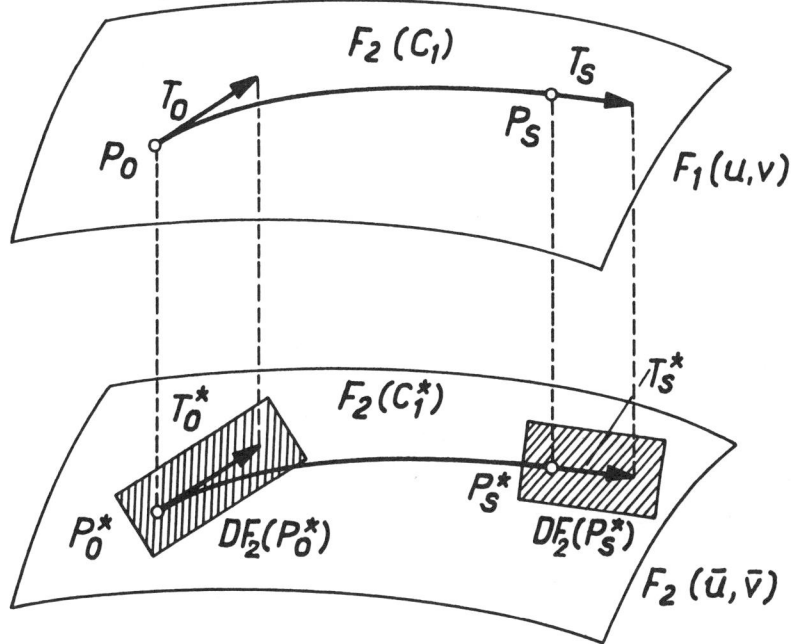

FIG. 3.23. *Geometric boundary conditions in the starting and ending point of the approximation curve.*

These tangent planes are determined by the partial derivatives

$$\left(\frac{\partial F_2}{\partial u}(\mathbf{P}_0^*), \frac{\partial F_2}{\partial v}(\mathbf{P}_0^*)\right) \quad \text{and} \quad \left(\frac{\partial F_2}{\partial u}(\mathbf{P}_s^*), \frac{\partial F_2}{\partial v}(\mathbf{P}_s^*)\right).$$

If we determine the components

$$\begin{pmatrix} \bar{u}_0^* \\ \bar{v}_0^* \end{pmatrix}, \begin{pmatrix} \bar{u}_s^* \\ \bar{v}_s^* \end{pmatrix} \quad \text{of } \mathbf{T}_0^*, \mathbf{T}_s^*$$

with respect to these derivatives, we get the starting tangent or end tangent of the curve C^* in the parametric domain. The lengths of the tangents are unknown and will be determined during the approximation process.

If the required curve has the parametric representation (3.29a) and the knot vector (3.29b), we obtain

$$\mathbf{d}_0^* = \begin{pmatrix} \bar{u}_0 \\ \bar{v}_0 \end{pmatrix}, \qquad \mathbf{d}_m^* = \begin{pmatrix} \bar{u}_s \\ \bar{v}_s \end{pmatrix},$$

$$\mathbf{d}_1^* = \begin{pmatrix} \bar{u}_0 \\ \bar{v}_0 \end{pmatrix} + \lambda_1 \left[\begin{pmatrix} \bar{u}_0^* \\ \bar{v}_0^* \end{pmatrix} - \begin{pmatrix} \bar{u}_0 \\ \bar{v}_0 \end{pmatrix} \right], \qquad \mathbf{d}_{m-1}^* = \begin{pmatrix} \bar{u}_s \\ \bar{v}_s \end{pmatrix} + \lambda_2 \left[\begin{pmatrix} \bar{u}_s \\ \bar{v}_s \end{pmatrix} - \begin{pmatrix} \bar{u}_s^* \\ \bar{v}_s^* \end{pmatrix} \right]$$

(3.31)

for the de Boor points at the end points of the curve (with $m := \bar{p}_i$).

Thus we have the scalar factors λ_1, λ_2 (with respect to (3.1)) and the (vector-valued) inner de Boor points $\mathbf{d}_2^*, \cdots, \mathbf{d}_{m-1}^*$ as variables of the optimization. Further, the knots of the knot vector are variable as variables for optimization. The whole approximation process runs through the following steps:

$$(3.32) \quad \text{objective function:} \quad \delta := \sum_{i=0}^{s} \left| \begin{pmatrix} \bar{u}_i \\ \bar{v}_i \end{pmatrix} - \sum_{j=0}^{m} \mathbf{d}_j^* N_{jk}(\bar{t}_i) \right|^2 \rightarrow \min .$$

1. Give an initial (uniform) knot vector $\mathbf{T}_{st} = (t_0, \cdots, t_{m+k})$. Set control index $M = 1$ with upper boundary M_0.

2. Reduce δ by optimization (the de Boor points $\mathbf{d}_j^*$ ($j = 2, \cdots, m-2$) and the scalar factors λ_1, λ_2 are used as variables).

2a. Determine initial values of λ_1, λ_2 with help of a cubic Bézier curve, which has G^1 continuity with the given curve in the boundary points (see (3.4a)) using (3.7a) and (3.8).

2b. Determine $\mathbf{d}_j^*$ with the least squares method and objective function (3.32), while λ_1, λ_2 are constant.

2c. Determine λ_1, λ_2 with the least squares method and the objective function (3.32), while $\mathbf{d}_j^*$ are constant.

3. Reduce δ by changing parameterization iteratively with a method introduced in [17], [20].

4. If total error $|\delta|$ is less than a given error ϵ_0 stop else $M = M + 1$, if $M \leq M_0$ goto 2b.

3.3.10. Results of Some Tests. We will demonstrate the effectiveness of the developed method with the help of some examples (see also [18]). We pick out two sets of patches of a wheel house of a car body (see Fig. 3.24a). SURF3 is an integral bicubic Bézier spline surface with two patches; SURF4 is an integral bicubic Bézier spline surface with 15 patches. Additionally in SURF3 we introduce two holes and an artificial boundary curve. The given curves are linear B-spline curves, where the crosses denote the boundary points of each B-spline curve. SURF3 is merged to one bicubic Bézier patch (maximum distance error 0.162 mm, maximum extension of the patch 120 mm). Figure 3.24b shows the given and the approximated trimmed surface and the corresponding parametric domain. The following error values are obtained: maximum distance between the given curve and the approximating surface is 0.24 mm; maximum distance between the boundary curves of the given and the approximating surface 0.33 mm; maximum distance between the given inner curves and the approximation is 0.11 mm. Figure 3.24c contains the given and the approximated trimmed surface SURF4 with the corresponding parametric domain. The 15 spline patches are merged to one bicubic Bézier patch, with the following error values: maximum distance between the surfaces 1.0 mm; maximum distance between the given curves and the approximating surface 0.57 mm; maximum distance between the boundary curves 0.57 mm (maximum extension of the patch 120 mm). As shown in Fig. 3.24a the two sets of Bézier patches have common boundary curves. But the two sets are not continuous; there is a maximal distance of 0.36 mm between corresponding points at these boundary curves. After the merging and the approximation process, this error grows to 0.59 mm. In general, such error values can be tolerated in car body industries.

Now we consider the approximate conversion of the quintic rational B-spline surface from Figs. 3.9a–c with (3 × 4) uniform B-spline segments into a set of (6 × 2) integral bicubic Bézier patches. Figure 3.25a contains the control net of the given surface. The weights at the two inner vertical rows of the surface control points are −0.1; the other weights are equal to 1. Figure 3.25b shows the new G^0 set of integral Bézier patches with a largest error of 0.38 mm (maximum extension of the patch 100 mm). A maximal error of the normal vectors of 4.6 degrees appears additionally. Figure 3.25c contains the approximation of a given hole described by one B-spline curve of degree 3. In the parametric domain, one can observe the effect of the parameter transformations. The maximal error between the given curve and the approximating surface is 0.29 mm; the maximal distance between the inner given and the approximating curve is 0.53 mm.

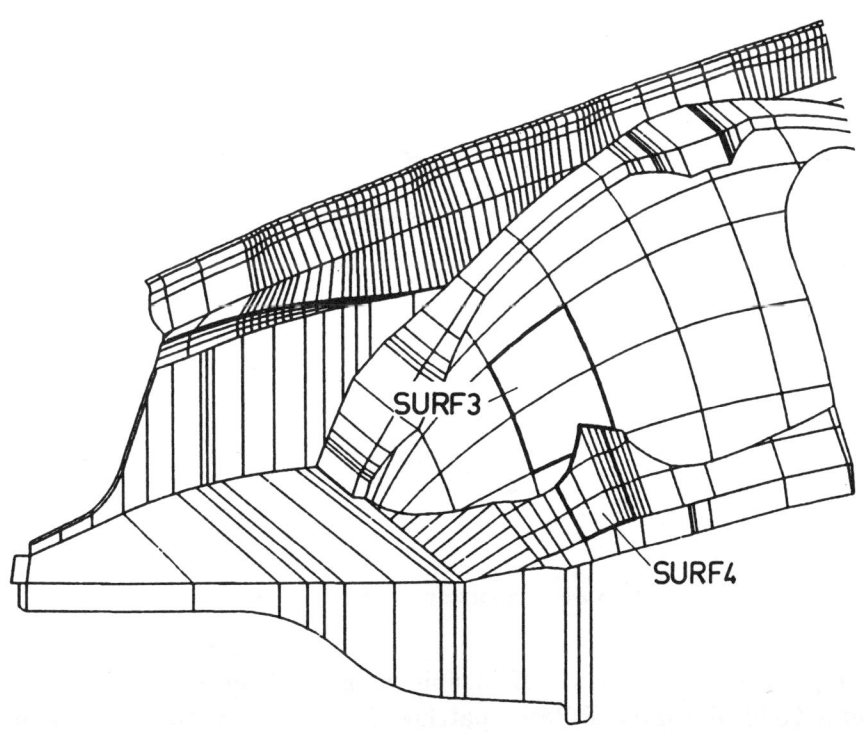

FIG. 3.24a. *Part of the wheel house of a car* (SURF3 *and* SURF4).

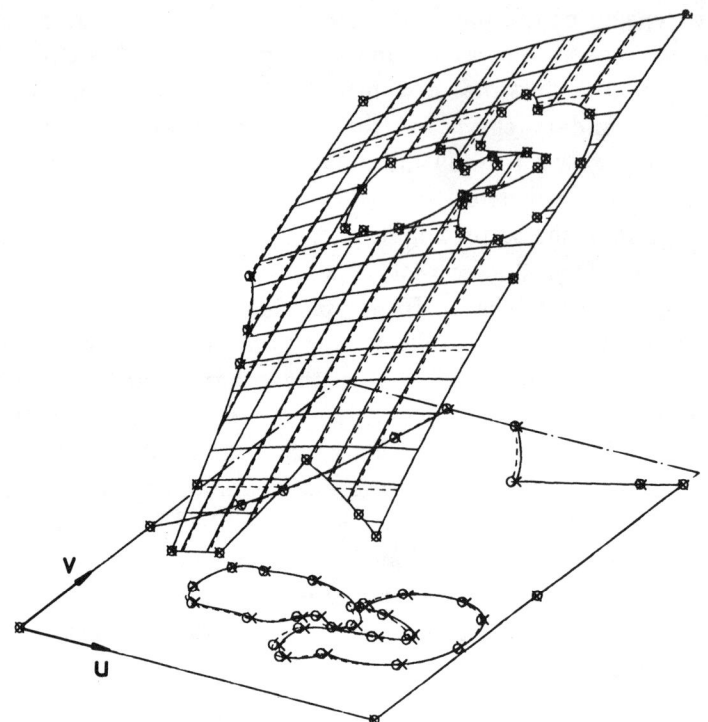

FIG. 3.24b. *Approximation of* SURF3 *with artificial holes.*

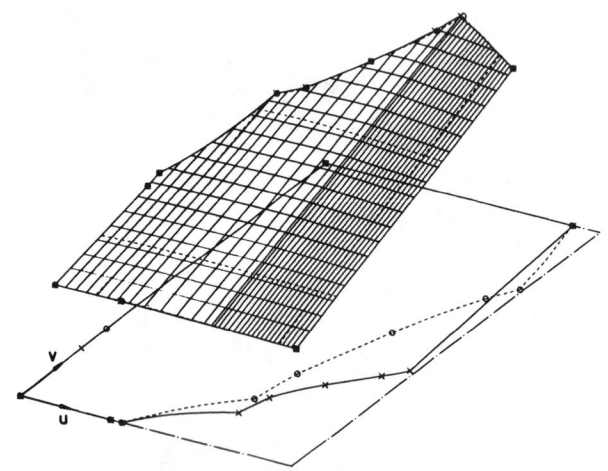

FIG. 3.24c. *Approximation of* SURF4.

In Figs. 3.26a,b a set of 10×22 bicubic integral Bézier patches is converted to a set of 28 bicubic integral Bézier patches. In Fig. 3.26b additionally a linear trimmed curve is merged to seven cubic segments. The given curve is drawn by solid lines; the approximating curve is drawn by dashed lines. One can observe the effect of the parameter correction in the corresponding parametric domain. The given curves are drawn by solid lines; the approximation is drawn

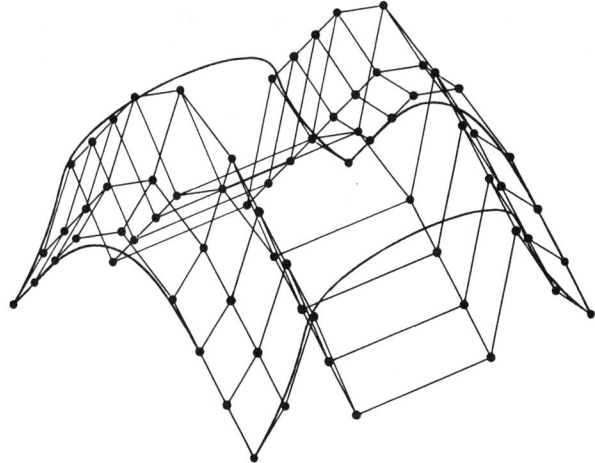

FIG. 3.25a. *Control net of B-spline surface from Fig. 3.9a,b.*

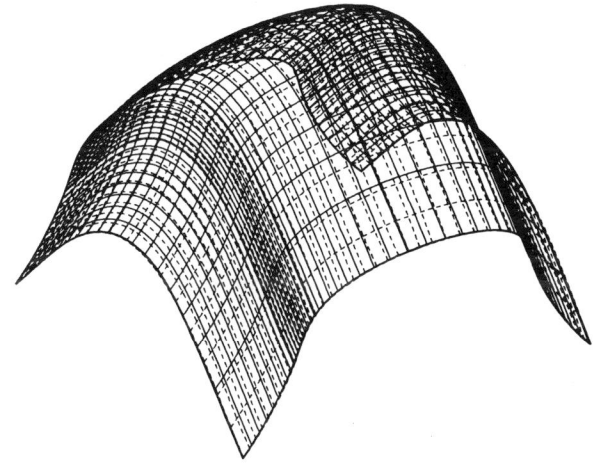

FIG. 3.25b. *Approximation by 6 × 2 patches.*

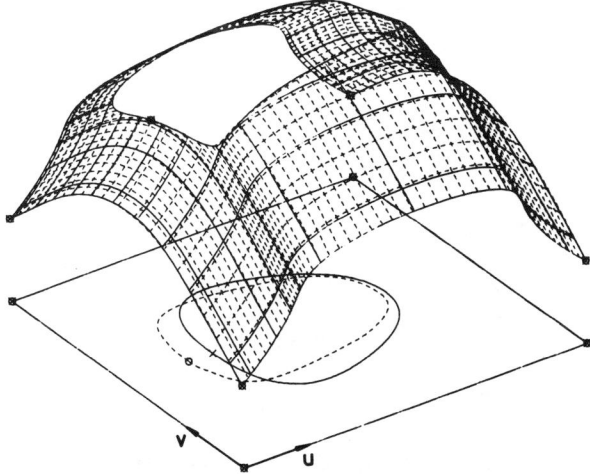

FIG. 3.25c. *Approximation of a hole on the surface from Fig. 3.25b.*

by dashed lines. On the surface the given and the approximating curves are drawn one on top of the other. The maximal extension of the set of surfaces is 80×30 cm. At the boundary curves, we get as maximal error 0.7 mm. In the interior, we get 0.6 mm. Maximal deviation of the trimmed curves is 0.8 mm. The following CPU time was used (on a VAX 8530): generic segmentation was 200 seconds; merging and additional segmentation depending on desired estimate error (less 1 mm) were 166 seconds; trimmed curve was 442 seconds.

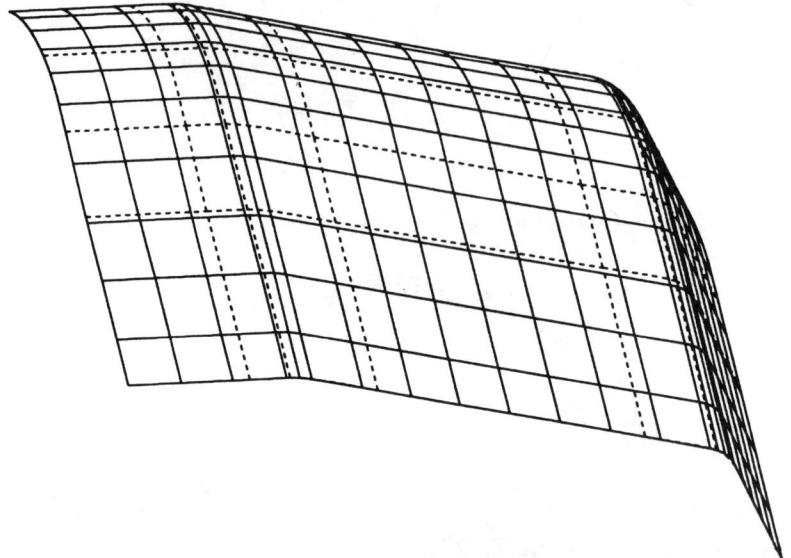

FIG. 3.26a. *Merging of 220 bicubic patches to 28 bicubic patches.*

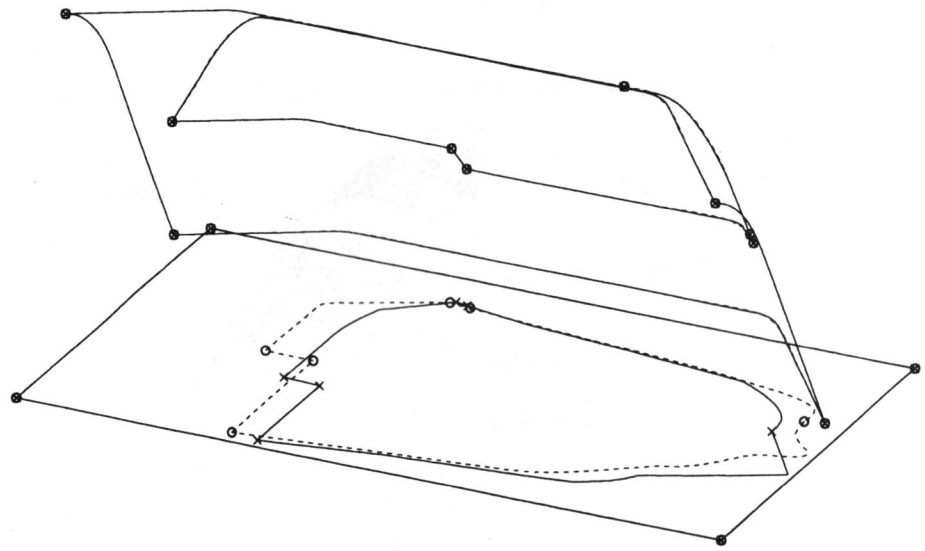

FIG. 3.26b. *Conversion of trimmed curves on surface out of Fig. 3.26a.*

3.4. Spline Approximation of Offset Curves and Offset Surfaces

3.4.1. Spline Approximation of Offset Curves. Offset curves have diverse potential applications in geometric modeling; for example, they are used for the generation of tool paths for numerical control machines; the description of the outer surface of a car body by an offset with material thickness; and, if the inner surface is the reference surface, for the definition of tolerance zones.

We suppose that the given curve has the parametric representation $\mathbf{X} = \mathbf{X}(t)$. Then the corresponding offset curve $\mathbf{X}_d$ at (oriented) distance d along the unit normal vector $\mathbf{n}(t)$ is given by

$$(3.33) \qquad\qquad \mathbf{X}_d(t) = \mathbf{X}(t) + \mathbf{n}(t)d.$$

For plane curves the normal vector $\mathbf{n}$ has the representation

$$(3.34) \qquad\qquad \mathbf{n}(t) = \frac{(-\dot{y}(t), \dot{x}(t))}{\sqrt{\dot{x}(t)^2 + \dot{y}(t)^2}}.$$

The normal vector (3.34) specifies a unique side of $\mathbf{X}$ on which the offset d is performed; the opposite side can be chosen by adopting a negative offset magnitude d.

We will assume that the given curve $\mathbf{X}(t)$ is a Bézier curve of degree n and that it has the parametric representation

$$(3.35) \qquad\qquad \mathbf{X}(t) = \sum_{i=0}^{n} \mathbf{V}_i B_i^n(t) \qquad (t \in [0,1]),$$

where $\mathbf{V}_i$ are the Bézier points of $\mathbf{X}(t)$. The corresponding offset curve $\mathbf{X}_d$ can be obtained by (3.33).

The approximation of the offset curve $\mathbf{X}_d$, for instance, will be either a nonrational Bézier curve of degree $m = 3$ or $m = 5$ (or a higher degree),

$$(3.36) \qquad\qquad \mathbf{Y}(t) = \sum_{i=0}^{m} \mathbf{W}_i B_i^m(t),$$

or a rational Bézier curve of degree $m = 3$,

$$(3.37) \qquad\qquad \mathbf{Y}(t) = \frac{\sum_{i=0}^{m} \beta_i \mathbf{W}_i B_i^m(t)}{\sum_{i=0}^{m} \beta_i B_i^m(t)},$$

where the control points $\mathbf{W}_i$ (and weights β_i) are the unknowns.

The approximation of the offset curve $\mathbf{X}_d$ by the approximating curve $\mathbf{Y}(t)$ can be performed in the following ways:

(i) *Segmentwise*, which means that one segment of the given curve $\mathbf{X}(t)$ can be approximated by one segment of the offset spline curve $\mathbf{Y}(t)$ corresponding to the offset curve $\mathbf{X}_d$ with respect to a given error tolerance ϵ_0;

(ii) *By splitting into further segments*, which means that one segment of the given curve $\mathbf{X}(t)$ leads to several spline segments of the approximating curve $\mathbf{Y}(t)$ of the offset curve $\mathbf{X}_d$ with respect to a given error tolerance ϵ_0;

(iii) *By merging of spline segments*, which means that several segments of the given spline curve will be approximated by one segment of the approximation $\mathbf{Y}(t)$ of the offset curve $\mathbf{X}_d$ with respect to a given error tolerance ϵ_0.

We will extend our geometric Hermite approximation methods developed in §§3.2, 3.3 on the *spline approximation* of offset curves and offset surfaces [16], [21], [20]. For cubic splines special approximating constructions of the offset curves are developed in [23] and [1]. In [9] approximations of special offset surfaces were given.

3.4.1.1. Approximation of offset curves by Bézier splines with G^1 continuity.

We suppose that the approximating curve $\mathbf{Y}(t)$ of the offset curve $\mathbf{X}_d$ has the degree $m = 3$, according to the set-up (3.36). Then the boundary points of the Bézier curve $\mathbf{Y}$ are determined by (see Fig. 3.27)

$$(3.38) \qquad \begin{aligned} \mathbf{W}_0 &= \mathbf{V}_0 + d\mathbf{n}(0), \\ \mathbf{W}_3 &= \mathbf{V}_n + d\mathbf{n}(1), \end{aligned}$$

with $\mathbf{n}(t)$ as unit normal vector of the given Bézier curve $\mathbf{X}(t)$. The first condition of (3.4a) leads further to

$$(3.39a) \qquad \mathbf{W}_1 = \mathbf{W}_0 + \lambda_1(\mathbf{V}_1 - \mathbf{V}_0),$$

$$(3.39b) \qquad \mathbf{W}_2 = \mathbf{W}_3 + \lambda_2(\mathbf{V}_n - \mathbf{V}_{n-1}),$$

where the λ_i are unknown quantities.

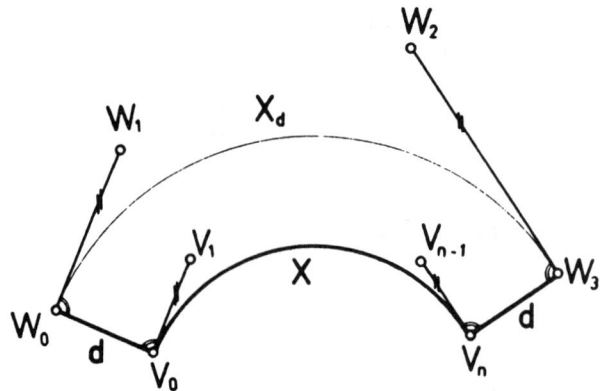

FIG. 3.27. *Boundary points of the approximation curve of the offset curve $\mathbf{X}_d$ of a given Bézier curve $\mathbf{X}$.*

To approximate the offset curve, we choose $(k + 1)$ equidistant points $\mathbf{P}_i$ on the given exact offset curve $\mathbf{X}_d$ with parameter values t_i $(i = 0(1)k)$ (with

$k > n$). Next, inserting these points $\mathbf{P}_i$ into the parametric representation of the approximation curve (3.36), we obtain

$$(3.40) \qquad\qquad \mathbf{P}_i = \sum_{j=0}^{3} \mathbf{W}_j B_j^3(t_i) + \boldsymbol{\delta}_i,$$

where $i = 0(1)k$ and $\boldsymbol{\delta}_i$ are the vectors of error. We minimize the total error sum according to §3.2.4.

Figure 3.28 illustrates an example of a (geometric) Bézier spline curve (see also [16]) with four cubic segments and two offset curves to both sides of the curve. The example was first considered in [1]. The Bézier points of the given curve are marked by solid boxes, the Bézier points of the approximation curve by open boxes. For the approximation of the offset curves, we need seven segments on both sides, as error value was chosen $\epsilon_0 = 0.01$ cm (with respect to the given scale). The cusps were evaluated by the algorithm described in [14]. The exact offset curves and their approximations are plotted one on top of the other.

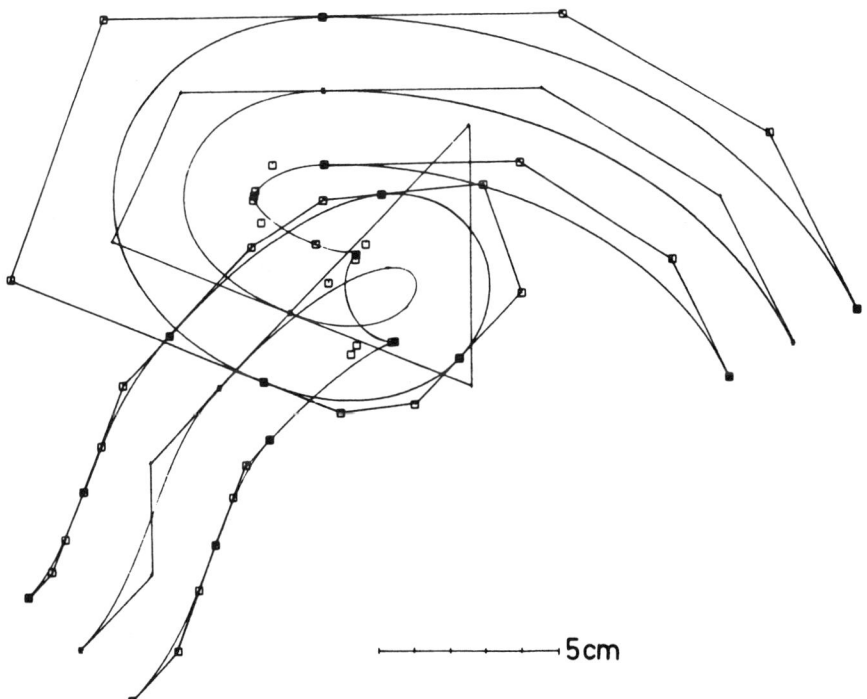

FIG. 3.28. *Bézier spline curve (degree 3) with seven segments (the intermediate curve with the small slope) and G^1 Bézier spline approximations (degree 3) of two offset curves (control polygons marked with boxes).*

3.4.1.2. Approximation of offset curves by Bézier splines with G^2 continuity. Next we suppose that the approximation curve $\mathbf{Y}(t)$ of the offset

curve, according to our set-up in (3.36), has the degree $m = 5$. The boundary points $\mathbf{W}_0$ and $\mathbf{W}_5$ of $\mathbf{Y}(t)$ are determined by (3.39a); the Bézier points $\mathbf{W}_1$ and $\mathbf{W}_4$ are determined as in (3.39b). Further we obtain from the G^2 condition (3.4b)

$$
\begin{aligned}
(3.41) \qquad \mathbf{W}_2 &= \mathbf{W}_0 + \lambda_1^2 k_0 (\mathbf{V}_2 - \mathbf{V}_1) + \mu_1 (\mathbf{V}_1 - \mathbf{V}_0), \\
\mathbf{W}_3 &= \mathbf{W}_5 + \lambda_2^2 k_1 (\mathbf{V}_{n-2} - \mathbf{V}_{n-1}) + \mu_2 (\mathbf{V}_n - \mathbf{V}_{n-1}),
\end{aligned}
$$

where μ_1, μ_2 are (arbitrary) quantities, and

$$
(3.42) \qquad k_i = \frac{m(n-1)}{n(m-1)} (1 + \kappa_i d)^{-1},
$$

and the κ_i are the curvatures at the boundary points $\mathbf{X}(i)$ with $i = 0, 1$ of the given curve $\mathbf{X}(t)$. The minimization of the total error sum follows analogously from §3.2.4.

Figure 3.29 contains the (quintic) approximation of a positive and negative offset curve of a given Bézier spline curve (two segments) of degree 19. The Bézier points of the given curve are marked by solid boxes; the Bézier points of the approximation curve are marked by open boxes. The error value of the upper curve is less than 0.014 cm; the error value of the lower curve is less than 0.003 cm. The exact offset curves and their approximations are plotted one on top of the other.

3.4.1.3. Approximation of offset curves by rational Bézier curves with G^1 continuity.
The approximation of an offset curve by rational splines leads in general to better results than the approximation by (ordinary) spline curves. If we use rational spline curves, more design parameters are available. The approximating curve may have the parametric equation (3.37), with $m = 3$ (see [16]). The given curve $\mathbf{X}(t)$ may have the Bézier representation (3.35). The exact offset curve $\mathbf{X}_d(t)$ is defined by (3.33). The geometric continuity conditions G^1 in (3.38) lead to the equations

$$
\begin{aligned}
(3.43) \qquad \mathbf{W}_1 &= \mathbf{W}_0 + \lambda_1 \frac{\beta_0}{\beta_1} (\mathbf{V}_1 - \mathbf{V}_0), \\
\mathbf{W}_2 &= \mathbf{W}_3 + \lambda_2 \frac{\beta_3}{\beta_2} (\mathbf{V}_n - \mathbf{V}_{n-1}).
\end{aligned}
$$

Analogously to §3.2, we choose equidistant points $\mathbf{P}_i$ on the exact offset curve $\mathbf{X}_d(t)$. Inserting these points into (3.37) and using (3.39a) and (3.45), one obtains the following equation:

$$
\begin{aligned}
(3.44) \qquad (\mathbf{P}_i - \mathbf{W}_0) B_0^3(t) &= \beta_1 (\mathbf{W}_0 - \mathbf{P}_i) B_1^3(t_i) + \beta_2 (\mathbf{W}_3 - \mathbf{P}_i) B_2^3(t_i) \\
&\quad + (\mathbf{W}_3 - \mathbf{P}_i) B_3^3(t_i) + \lambda_1 (\mathbf{V}_1 - \mathbf{V}_0) B_1^3(t_i) \\
&\quad + \lambda_2 (\mathbf{V}_n - \mathbf{V}_{n-1}) B_2^3(t_i) + \boldsymbol{\delta}_i,
\end{aligned}
$$

with $\boldsymbol{\delta}_i$ as error vector. The total error is a function of four variables: β_1, β_2, λ_1, λ_2. We can determine the unknowns by the least square methods according to

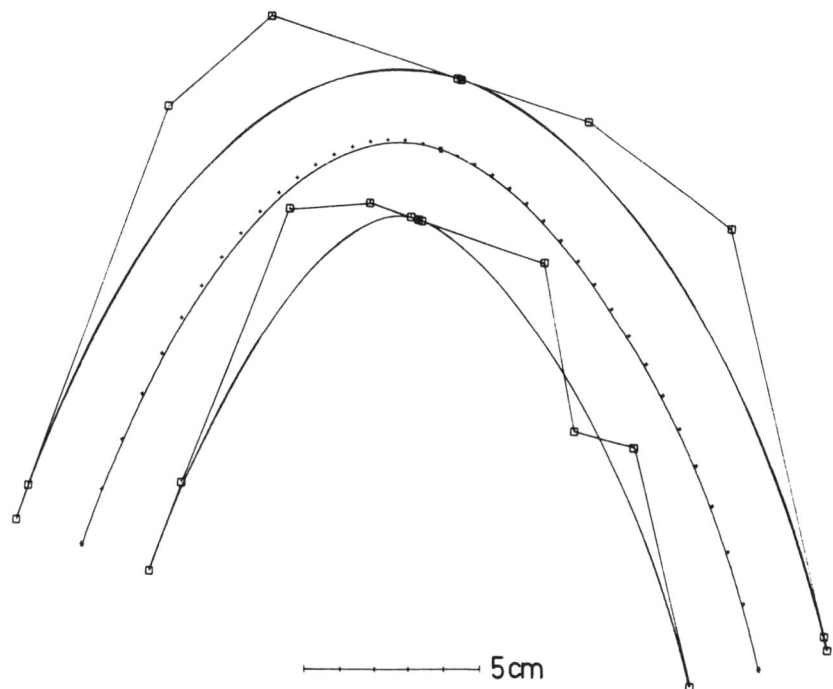

FIG. 3.29. *Bézier spline curve (degree 19) with two segments (the intermediate
curve) and G^2 Bézier spline approximation (degree 5) of two offset curves.*

§3.2 and continue as described in the algorithm for G^1 approximation of offset
curves. If we approximated the offset curves from Fig. 3.28 by rational cubic
Bézier splines, we would have needed only four segments for the offset curve
without cusps, and we would have obtained only five segments for the other
offset curve.

3.4.2. Spline Approximation of Offset Surfaces.   The techniques used
can be extended to spline approximation of offset surfaces [19]. We suppose
that the given surface is a tensor product Bézier surface of degree (n, m) and
that it has a parametric representation $\mathbf{X} = \mathbf{X}(u, v)$. Then the corresponding
offset surface $\mathbf{X}_d$ at (oriented) distance d along the unit normal vector $\mathbf{N}(u, v)$
is given by

$$(3.45) \qquad \mathbf{X}_d(u, v) = \mathbf{X}(u, v) + \mathbf{N}(u, v)d.$$

$\mathbf{N}$ specifies a unique side of $\mathbf{X}$ on which the offset d is performed. The opposite
side can be chosen by adopting a negative offset magnitude. If $\mathbf{X}(u, v)$ has the
parametric representation

$$\mathbf{X}(u, v) = \sum_{i=0}^{n} \sum_{k=0}^{m} \mathbf{V}_{ik} B_i^n(u) B_k^m(v),$$

then the Bézier approximation of the offset surface may have the parametric representation

$$\mathbf{Y}(u,v) = \sum_{i=0}^{p}\sum_{k=0}^{q} \mathbf{W}_{ik}B_i^p(u)B_k^q(v).$$

The corners of the offset surface are determined by

$$(3.46) \quad \begin{aligned} \mathbf{W}_{00} &= \mathbf{V}_{00} + \mathbf{N}(0,0)d, \\ \mathbf{W}_{0q} &= \mathbf{V}_{0m} + \mathbf{N}(0,1)d, \\ \mathbf{W}_{p0} &= \mathbf{V}_{n0} + \mathbf{N}(1,0)d, \\ \mathbf{W}_{pq} &= \mathbf{V}_{nm} + \mathbf{N}(1,1)d. \end{aligned}$$

Analogously to §3.3 we suppose that the offset surface $\mathbf{X}_d$ and the approximating surface $\mathbf{Y}$ will have (1) contact of order 1 in the bicubic case, and (2) contact of order 2 in the biquintic case, in the corners. So we have the same situation described in §3.3 and we can proceed in the approximating process as described in that section. Figure 3.30 shows an example of a given (4, 4) Bézier surface and the (3, 3) Bézier approximation of the two offset surfaces with different oriented offsets d_1 and d_2.

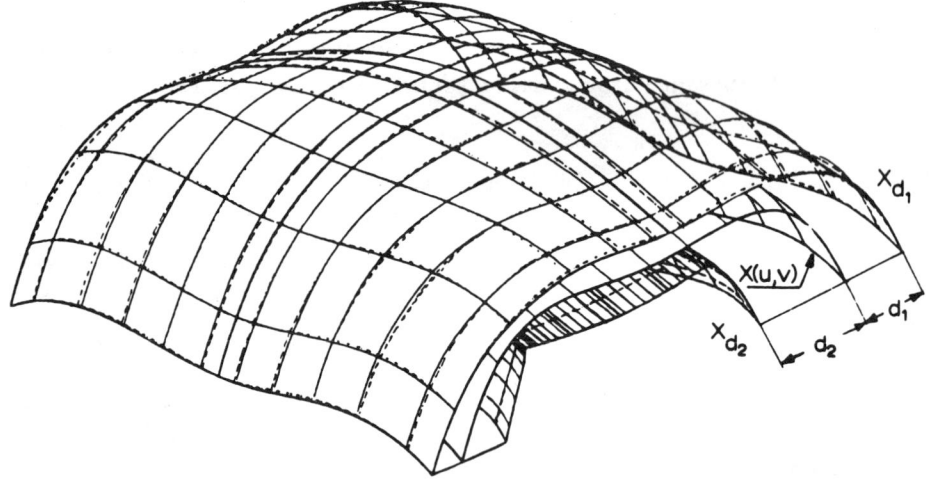

FIG. 3.30.   *Spline approximation of offset surfaces by bicubic patches (given surfaces $\mathbf{X}(u)$; offset surfaces with offsets d_1, d_2).*

To reduce the approximation error, the two approximating surfaces of the two given offset surfaces are subdivided: $\mathbf{X}_{d_2}$ is segmented into eight patches, while $\mathbf{X}_{d_1}$ is subdivided into only four patches. Thus a large approximation error appears, as can be seen at the boundary curves. If $\mathbf{X}_{d_1}$ were segmented into eight patches, these large errors would vanish.

References

[1] R. Arnold, *Quadratische und kubische Offset-Bézierkurven*, Ph.D. thesis, Universität Dortmund, Germany, 1986.

[2] B. A. Barsky, *Computer Graphics and Geometric Modeling Using β-Splines*, Springer-Verlag, Berlin, New York, 1988.

[3] L. Bardis and N. M. Patrikalakis, *Approximate conversion of rational splines*, Comput. Aided Geom. Des., 6 (1989), pp. 189–204.

[4] J. Brode, *Konvertieren von Polynomen in CAGD*, Diplom-Arbeit, TU Braunschweig, Germany, 1990.

[5] S. Cohen, *Beitrag zur steuerbaren Interpolation von Kurven und Flächen*, Ph.D. thesis, TU Dresden, Germany, 1982.

[6] C. de Boor, *A Practical Guide to Splines*, Springer-Verlag, Berlin, New York, 1978.

[7] L. Dannenberg and H. Nowacki, *Approximate conversion of surface representations with polynomial bases*, Comput. Aided Geom. Des., 2 (1985), pp. 123–132.

[8] G. Farin, *Curves and Surfaces for Computer Aided Geometric Design. A Practical Guide*, Academic Press, New York, 1988.

[9] R. T. Farouki, *The approximation of non-degenerated offset surfaces*, Comput. Aided Geom. Des., 3 (1986), pp. 15–43.

[10] R. T. Farouki and V. T. Rajan, *Algorithms for polynomials in Bernstein form*, Comput. Aided Geom. Des., 5 (1988), pp. 1–26.

[11] G. Geise, *Über berührende Kegelschnitte einer ebenen Kurve*, Z. Angew. Math., 42 (1962), pp. 297–304.

[12] J. Hoschek and D. Lasser, *Grundlagen der geometrischen Datenverarbeitung*, Teubner, Stuttgart, 1989.

[13] G. Hölzle, *Knot placement for piecewise polynomial approximation of curves*, Comput. Aided Des., 15 (1983), pp. 295–296.

[14] J. Hoschek, *Offset curves in the plane*, Comput. Aided Des., 17 (1985), pp. 77–82.

[15] _____, *Approximate conversion of spline curves*, Comput. Aided Geom. Des., 4 (1987), pp. 59–66.

[16] _____, *Spline approximation of offset curves*, Comput. Aided Geom. Des., 5 (1988), pp. 33–40.

[17] _____, *Intrinsic parameterization for approximation*, Comput. Aided Geom. Des., 5 (1988), pp. 27–31.

[18] J. Hoschek and F. J. Schneider, *Spline conversion for trimmed rational Bézier and B-spline surfaces*, Comput. Aided Des., accepted for publication.

[19] _____, *Spline approximation of offset curves and offset surfaces*, in Proceedings of the Third Conference Mathematics in Industry, J. Manley, et al., ed., Kluwer-Teubner, Stuttgart, 1990, pp. 383–389.

[20] J. Hoschek, F. J. Schneider, and P. Wassum, *Optimal approximate conversion of spline surfaces*, Comput. Aided Geom. Des., 6 (1989), pp. 293–306.

[21] J. Hoschek and N. Wissel, *Optimal approximate conversion of spline curves and spline approximation of offset curves*, Comput. Aided Des., 20 (1988), pp. 457–483.

[22] M. Kallay, *Approximating a composite cubic curve by one with fewer pieces*, Comput. Aided Des., 19 (1987), pp. 539–543.

[23] R. Klass, *An offset spline approximation for plane cubic splines*, Comput. Aided Des., 15 (1983), pp. 297–299.

[24] M. A. Lachance, *Chebyshev economization for parametric surfaces*, Comput. Aided Geom. Des., 5 (1988), pp. 195–208.

[25] N. M. Patrikalakis, *Approximate conversion of rational splines*, Comput. Aided Geom. Des., 6 (1989), pp. 155–166.

[26] B. Su and D. Liu, *Computational Geometry*, Academic Press, New York, 1989.

[27] P. Wassum, *Approximative Basistransformation von Spline-Flächen mit beliebigem Polynomgrad*, Fachbereich Mathematik, Technische Hochschule, Darmstadt, Germany, 1988, preprint.

[28] _____, *Bedingungen und Konstruktionen zur geometrischen Stetigkeit und Anwendungen auf approximative Basistransformationen*, Ph.D. thesis, Technische Hochschule, Darmstadt, Germany, 1991.

[29] M. A. Watkins and A. J. Worsey, *Degree reduction of Bézier curves*, Comput. Aided Des., 20 (1988), pp. 398–405.

Degree Reduction Fairing of Cubic B-Spline Curves

Gerald Farin

4.1. Basics

Cubic B-splines have established themselves as an indispensable tool for the design of curves in many applications. This paper addresses the problem of generating aesthetically pleasing, or fair, curves in the B-spline form.

For a detailed description of cubic B-splines, as well as their conversion to the piecewise Bézier form, see [7]. A cubic B-spline curve is defined by a knot sequence $u_0 < \cdots < u_L$ and a control polygon $\mathbf{d}_{-1}, \cdots, \mathbf{d}_{L+1}$. The B-spline curve is C^2 overall and a cubic curve over each interval $[u_i, u_{i+1}]$. This cubic has a Bézier polygon $\mathbf{b}_{3i}, \mathbf{b}_{3i+1}, \mathbf{b}_{3i+2}, \mathbf{b}_{3i+3}$, found from

$$(4.1) \qquad \mathbf{b}_{3i} = \frac{\Delta_i}{\Delta_{i-1} + \Delta_i}\mathbf{b}_{3i-1} + \frac{\Delta_{i-1}}{\Delta_{i-1} + \Delta_i}\mathbf{b}_{3i+1},$$

where

$$(4.2) \qquad \mathbf{b}_{3i+1} = \frac{\Delta_i + \Delta_{i+1}}{\Delta}\mathbf{d}_i + \frac{\Delta_{i-1}}{\Delta}\mathbf{d}_{i+1},$$

$$(4.3) \qquad \mathbf{b}_{3i+2} = \frac{\Delta_{i+1}}{\Delta}\mathbf{d}_i + \frac{\Delta_{i-1} + \Delta_i}{\Delta}\mathbf{d}_{i+1},$$

with

$$(4.4) \qquad \Delta = \Delta_{i-1} + \Delta_i + \Delta_{i+1}$$

and $\Delta_i = u_{i+1} - u_i$.

4.2. Shape and Curvature

It is known from differential geometry that a planar curve is uniquely defined by its curvature [6]. We may visualize curvature by plotting curvature versus arc length, generating the so-called *curvature plot*.[1] The curvature plot is an extremely accurate tool in providing information about the shape of a curve.

---

[1] It is computationally easier to plot curvature versus the curve parameter. This is usually sufficient, but since the parameterization of a curve is not unique, the same curve may have different curvature plots when plotted against the different parameters.

In fact, two curves may not be visually distinguishable due to poor screen resolution or size, and yet may exhibit quite different curvature plots (see the examples below). These differences, subtle as they may be, may be crucial in an environment where aesthetics or aerodynamics are important, i.e., in the automotive, aircraft, or naval industries.

To a practitioner in the field, it is obvious that the analysis tool "curvature plot" is essential, but this insight is only slowly gaining acceptance in the academic community. In fact, almost all definitions of the (aesthetic) fairness of a curve rely on the notion of a curvature plot. We use the following:

A curve is fair if its curvature plot
consists of relatively few monotone pieces.

Sabin has suggested that "a frequency analysis of the radius of curvature plotted against arc length might give some measure of fairness — the lower the dominant frequency, the fairer the curve" (quoted from Forrest [10]). It also appears that these definitions of fairness are linked to the *information content* present in the curvature plot: the lower the information content, the fairer the curve. It is this interpretation of the definition of fairness that the subsequent algorithms will utilize.

Our definition of fairness can also be found in [5], [3], and [26]. In most cases, one would like the curvature to be continuous. Sometimes, as recently pointed out to me by Bézier, a discontinuity in curvature is admissible if there is no discontinuity of the *slope* of the curvature at the point in question.

I would like to add a personal note. There are definitions of curve fairness that do not address the shape of the curvature plot. The most common one is: *a curve is fair if it minimizes $\int [\kappa(s)]^2 ds$,* where $\kappa(s)$ denotes curvature as a function of arc length. This minimum property is satisfied by the elastic beam. I fail to see what an elastic beam has to do with aesthetic appearance. The reason for the minimum property definition of fairness is that it lends itself to mathematical treatment, in particular when used in the austere version that replaces curvature by a second derivative (see Lee [18] for a more in depth discussion of this subject).

In summary, the above definition was not arrived at through a desire for mathematical tractability, but by empirically quantifying what designers regard as aesthetic (fair). Thus we hope that the resulting methods will also be close to the intentions of a designer.

4.3. Knot Removal Fairing

A typical problem in the design process is that of *digitizing errors*: data points have been obtained from some digitizing device (a tablet being the simplest), and a fair curve is sought through them. In many cases, the digitized data are inaccurate, and this presence of digitizing error manifests itself in a "rough" curvature plot of an interpolating spline curve.[2] Splines that are obtained from interactive adjustment of control polygons usually exhibit rough curvature plots as well.

In [8], [9], and [25], we have presented algorithms that fair B-spline curves in the sense that they improve their curvature plots. In this paper, those algorithms are referred to as *knot removal fairing algorithms*. Specifically, those algorithms reduce the discontinuities in the slope of curvature between adjacent cubic pieces in the following way. The control point $\mathbf{d}_i$ of a C^2 cubic B-spline curve is associated with the parameter value u_i. If the B-spline curve were three times differentiable at u_i instead of just twice, it would not have a discontinuity in the slope of its curvature.[3] Thus we try to move $\mathbf{d}_i$ to a new position $\hat{\mathbf{d}}_i$ such that the new curve is now C^3 at u_i.

After some calculation (equating the left and the right third derivative of the new spline curve), one verifies that the new vertex $\hat{\mathbf{d}}_i$ is given by

$$(4.5) \qquad \hat{\mathbf{d}}_i = \frac{(u_{i+2} - u_i)\mathbf{l}_i + (u_i - u_{i-2})\mathbf{r}_i}{u_{i+2} - u_{i-2}},$$

where the auxiliary points $\mathbf{l}_i, \mathbf{r}_i$ are given by

$$\mathbf{l}_i = \frac{(u_{i+1} - u_{i-3})\mathbf{d}_{i-1} - (u_{i+1} - u_i)\mathbf{d}_{i-2}}{u_i - u_{i-3}}$$

and

$$\mathbf{r}_i = \frac{(u_{i+3} - u_{i-1})\mathbf{d}_{i+1} - (u_i - u_{i-1})\mathbf{d}_{i+2}}{u_{i+3} - u_i}.$$

In practice, the improved vertex $\hat{\mathbf{d}}_i$ may be further away from the original vertex $\mathbf{d}_i$ than a prescribed tolerance allows. In that case, one restricts a realistic $\hat{\mathbf{d}}_i$ to be in the direction towards the optimal $\hat{\mathbf{d}}_i$, but within tolerance to the old $\mathbf{d}_i$.

Knot removal fairing can be made more sophisticated than described here; for instance, Sapidis devised an algorithm to enforce convexity in the fairing process (see [24] and [25]).

Other methods for curve fairing exist. We mention Kjellander's method [16] which moves a data point to a more favorable location and then interpolates the changed data set with a C^2 cubic spline. This method is global. A method that fairs only data points, not spline curves, is presented by Renz [23]. Methods that aim at the smoothing of single Bézier curves are discussed by Hoschek [12], [13]. An algorithm by MacCallum and Zhang [19] smoothes a uniform cubic B-spline curve by reinterpreting the control polygon as that of a quartic one. Since that method is both global and only works for uniform knot spacing, it was not considered here.

---

[2]One may argue that the rough curvature plot is merely the artifact of an inadequate interpolation method. But a typical effect of inaccurate data is that convexity is destroyed — and that we thus observe "wiggles" in the curvature plot. *No* interpolation method will produce a convex curve through such perturbed data points.

[3]This statement is only true for nondegenerate curves. An exceptional case is pointed out by Lee [18].

4.4. Degree Reduction Fairing

The idea behind the algorithm (4.5) is to equate third derivatives from the left and right at u_i. If the Bézier polygons of the two involved cubics are denoted by $\mathbf{b}_{3i-3}, \mathbf{b}_{3i-2}, \mathbf{b}_{3i-1}, \mathbf{b}_{3i}$ and $\mathbf{b}_{3i}, \mathbf{b}_{3i+1}, \mathbf{b}_{3i+2}, \mathbf{b}_{3i+3}$, the condition for equal third derivatives at u_i becomes (see [7])

$$(4.6) \qquad \frac{\Delta^3 \mathbf{b}_{3i-3}}{(\Delta_{i-1})^3} = \frac{\Delta^3 \mathbf{b}_{3i}}{(\Delta_i)^3},$$

where the Δ^3 denote third forward differences and again $\Delta_i = u_{i+1} - u_i$.

The idea behind *degree reduction fairing* is not to equate third differences, as in (4.6), but rather to make each individual third difference $\Delta^3 \mathbf{b}_{3i}$ small. Recall that

$$(4.7) \qquad \Delta^3 \mathbf{b}_{3i} = \mathbf{b}_{3i} - 3\mathbf{b}_{3i+1} + 3\mathbf{b}_{3i+2} - \mathbf{b}_{3i+3} = \mathbf{0}$$

means that the cubic defined by $\mathbf{b}_{3i}, \mathbf{b}_{3i+1}, \mathbf{b}_{3i+2}, \mathbf{b}_{3i+3}$ is actually a quadratic curve, i.e., a parabola. It seems reasonable to hope for a better curve shape if all cubic pieces are close to being parabolic. The rationale is that parabolas are simpler in shape (i.e., have less information content) than cubics, and when we try to combat digitizing errors, we also combat an overabundance of information in the curve shape.

We could rewrite (4.7) in terms of the B-spline coefficients $\mathbf{d}_i$ according to (4.1)–(4.3). This would lead to a global linear system for new control points $\hat{\mathbf{d}}_i$. There is a simpler geometric method that achieves the same goal. In addition, it will have the property of being local.

A cubic may be approximated by a quadratic by the process of *degree reduction*; see [11], [7], [27]. The approximating quadratic, with Bézier points $\mathbf{c}_{2i}, \mathbf{c}_{2i+1}, \mathbf{c}_{2i+2}$, may be obtained as

$$\mathbf{c}_{2i} = \mathbf{b}_{3i}, \quad \mathbf{c}_{2i+1} = \mathbf{m}_i, \quad \mathbf{c}_{2i+2} = \mathbf{b}_{3i+3},$$

where

$$(4.8) \qquad \mathbf{m}_i = \frac{1}{2}\left(\frac{3}{2}\mathbf{b}_{3i+1} - \frac{1}{2}\mathbf{b}_{3i}\right) + \frac{1}{2}\left(\frac{3}{2}\mathbf{b}_{3i+2} - \frac{1}{2}\mathbf{b}_{3i+3}\right).$$

Other meaningful ways to define $\mathbf{m}_i$ may exist, but this was not explored.

The quadratic thus defined may be brought back to cubic form by the process of *degree elevation*.

The involved Bézier points may be expressed in terms of the B-spline vertices $\mathbf{d}_{i-1}, \mathbf{d}_i, \mathbf{d}_{i+1}, \mathbf{d}_{i+2}$ according to the formulas given above. Let us agree to keep $\mathbf{d}_{i-1}$ and $\mathbf{d}_{i+2}$ fixed and to only change the other two control vertices. This simplification leads to the following algorithm to fair the *i*th curve segment.

ALGORITHM.

step 1 Given the initial B-spline curve, compute the Bézier points $\mathbf{b}_{3i}$, $\mathbf{b}_{3i+1}, \mathbf{b}_{3i+2}, \mathbf{b}_{3i+3}$.

step 2 Compute the quadratic approximation to the cubic defined by these Bézier points.

step 3 Degree elevate that quadratic, resulting in a cubic control polygon $\mathbf{b}_{3i}, \hat{\mathbf{b}}_{3i+1}, \hat{\mathbf{b}}_{3i+2}, \mathbf{b}_{3i+3}$.

step 4 The new $\hat{\mathbf{d}}_i$ and $\hat{\mathbf{d}}_{i+1}$ are on the straight line through $\hat{\mathbf{b}}_{3i+1}, \hat{\mathbf{b}}_{3i+2}$. The involved ratios are known from (4.2) and (4.3). They are illustrated in Fig. 4.1.

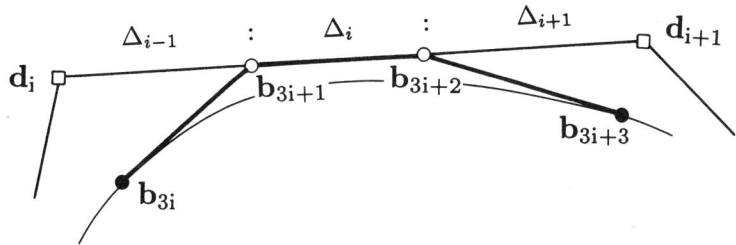

FIG. 4.1. *Cubic B-spline curves. The relationship between B-spline control polygon and Bézier control points.*

We now give the formulas for each step.

ALGORITHM (Formulas).
step 1 The necessary formulas are given by (4.1)–(4.3) in §4.1.
step 2 This involves the computation of $\mathbf{m}_i$, which is given in (4.8).
step 3 We obtain $\hat{\mathbf{b}}_{3i+1}, \hat{\mathbf{b}}_{3i+2}$ from

$$\hat{\mathbf{b}}_{3i+1} = \frac{2\mathbf{m}_i + \mathbf{b}_{3i}}{3}$$

and

$$\hat{\mathbf{b}}_{3i+2} = \frac{2\mathbf{m}_i + \mathbf{b}_{3i+3}}{3}.$$

step 4 The desired new control points $\hat{\mathbf{d}}_i$ and $\hat{\mathbf{d}}_{i+1}$ are given by

(4.9)
$$\hat{\mathbf{d}}_i = \frac{(\Delta_{i-1} + \Delta_i)\hat{\mathbf{b}}_{3i+1} - \Delta_{i-1}\hat{\mathbf{b}}_{3i+2}}{\Delta_i}$$

and

$$\hat{\mathbf{d}}_{i+1} = \frac{(\Delta_i + \Delta_{i+1})\hat{\mathbf{b}}_{3i+2} - \Delta_{i+1}\hat{\mathbf{b}}_{3i+1}}{\Delta_i},$$

where i ranges from 0 to L. In order to avoid problems for $i = 0$ and $i = L$, we introduce additional knots $u_1 = u_0$ and $u_{L+1} = u_L$. The control points $\mathbf{d}_{-1}$ and $\mathbf{d}_{L+1}$, being the first and last points of the curve, are not changed.

step 5 The $\hat{\mathbf{d}}_i$ in (4.9) was already computed as $\hat{\mathbf{d}}_{i+1}$ for the previous i. We average the result from (4.9) with the previous value:

(4.10)
$$\hat{\mathbf{d}}_i = \frac{1}{2}\frac{(\Delta_{i-1} + \Delta_i)\hat{\mathbf{b}}_{3i+1} - \Delta_{i-1}\hat{\mathbf{b}}_{3i+2}}{\Delta_i} + \frac{1}{2}\hat{\mathbf{d}}_i.$$

This is not the only way to avoid a double definition of control points — we have obtained good results using it.

It should be kept in mind that the above algorithm does not produce a curve which consists of quadratic (i.e., degree reduced) segments only:

(a) only d_i and d_{i+1} are changed, whereas also d_{i-1} and d_{i+2} have to be changed in order to build a degree reduced cubic; and

(b) the result of the $(i-1)$st step is distorted in the ith step by (4.10).

It may be argued that (b) takes away from the idea of degree reducing each curve segment. This is true, and I presently know only one argument to explain why (b) does not pose a serious problem: the method works!

4.5. Comparison

Before we describe differences between knot removal fairing and degree reduction fairing, it is interesting to point to a common theme to which both adhere. Both methods aim at a simplification in curve shape, or at a removal of superfluous information. Removing a knot from a B-spline curve changes its information content, and so does the reduction of the polynomial degree of a segment. So both schemes have the same high level description: Remove information from the curve by suitably eliminating some of its degrees of freedom, and then formally rewrite it to be compatible with the initial curve representation.

Both knot removal fairing (4.5) and the new degree reduction methods were implemented and tested on data that are a reasonable simulation of real-world design data.

Although the methods are local, we applied them to every segment of a given spline curve, assuming that all B-spline coefficients were in error as the result of digitizing. An automatic localization for knot removal smoothing has been implemented by Sapidis [24]. It looks for the knot where two cubics join with the largest discontinuity in the slope of the curvature plot and then fairs only there.

An automatic localization for degree reduction fairing might look for the cubic segment with the largest third derivative, and then fair only there.

In the following examples, illustrated in Figs. 4.3–4.8, we compare degree reduction fairing to knot removal fairing. In all examples, we did not impose a limit on the maximal deviation between the old and new curve, in order to see what each algorithm deems the most appropriate change. In practice, such limits will be necessary. In most examples, it was not possible to tell the curves apart as displayed on the screen; their plots are omitted.

The following observations were made.

Magnitude of change. Knot removal fairing changes a curve more than does degree reduction fairing. The deviation was measured by simply (and somewhat naively) finding the largest distance $\|d_i - \hat{d}_i\|$.

FIG. 4.2. *Cusps. Two cubic segments are shown that have a cusp (left). Knot removal fairing produces no change, whereas degree reduction fairing produces the right curve after seven iterations.*

FIG. 4.3. *Original curvature plot.*

Fairing speed. Typically, one has to fair a curve several times until the curvature plot (which improves with each iteration) looks satisfactory. Knot removal fairing needs about two to three iterations, while degree reduction fairing needs up to five iterations. Even so, degree reduction fairing changes the curve less.

Cusps. It has been pointed out by Lee [18] that knot removal fairing will tolerate cusps at the segment end points. Degree reduction fairing does not have this disadvantage, as shown in Fig. 4.2. (It should be kept in mind that this example is of theoretical, not of practical, relevance.)

Dimensionality. Both methods may be applied to space curves, but this has not been tested.

4.6. Surfaces

Once imperfections are detected in a tensor product B-spline surface (see [4], [7]), one would want methods to remove them without time-consuming interactive adjustment of control polygons.

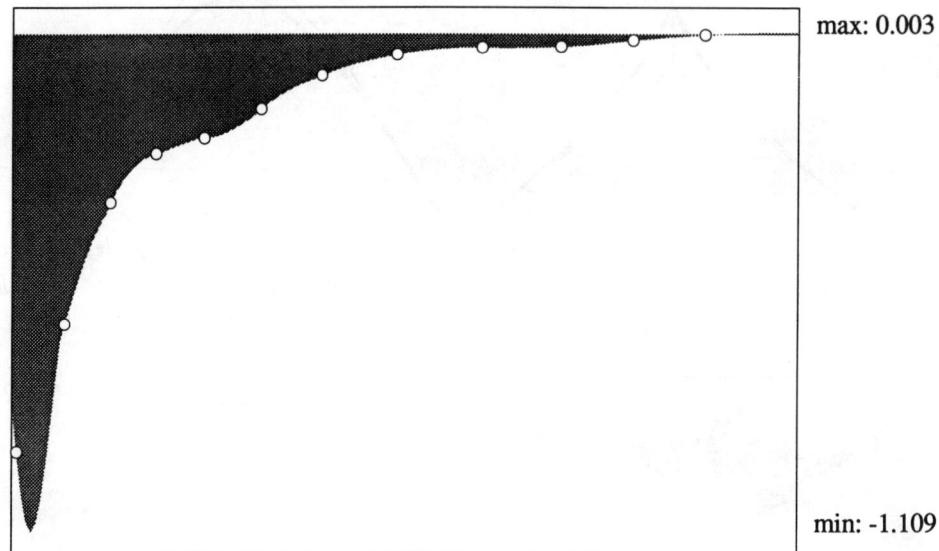

FIG. 4.4. *Curvature plot after two applications of knot removal fairing. Deviation from original polygon: 0.015.*

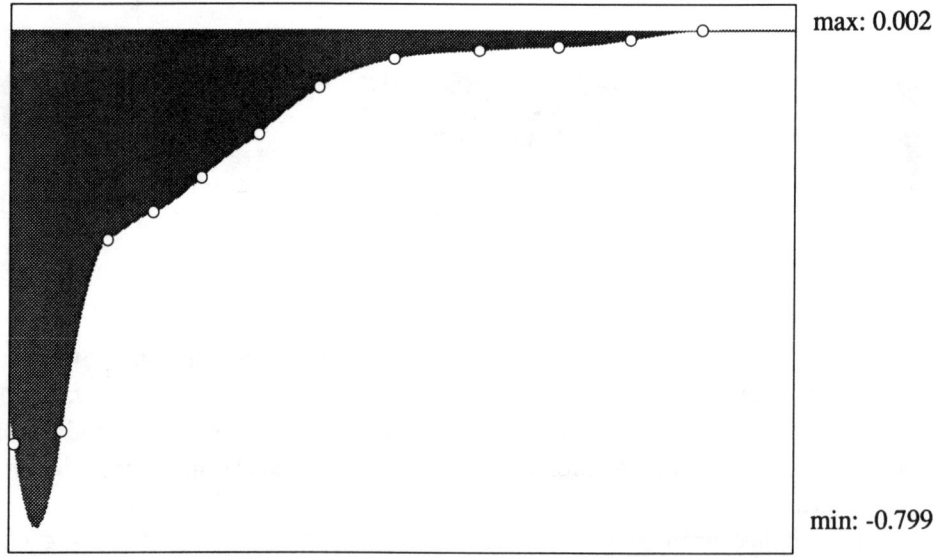

FIG. 4.5. *After four applications of degree reduction fairing. Deviation from original polygon: 0.006.*

The following algorithm fairs a tensor product B-spline surface.

ALGORITHM.

Let $\{\mathbf{d}_{i,j}\}$ be the control net of that surface with knot sequences $\{u_i\}$ and $\{v_j\}$. Fair the surface by first interpreting all rows of the control net as B-spline control polygons and then applying the curve fairing algorithm to each

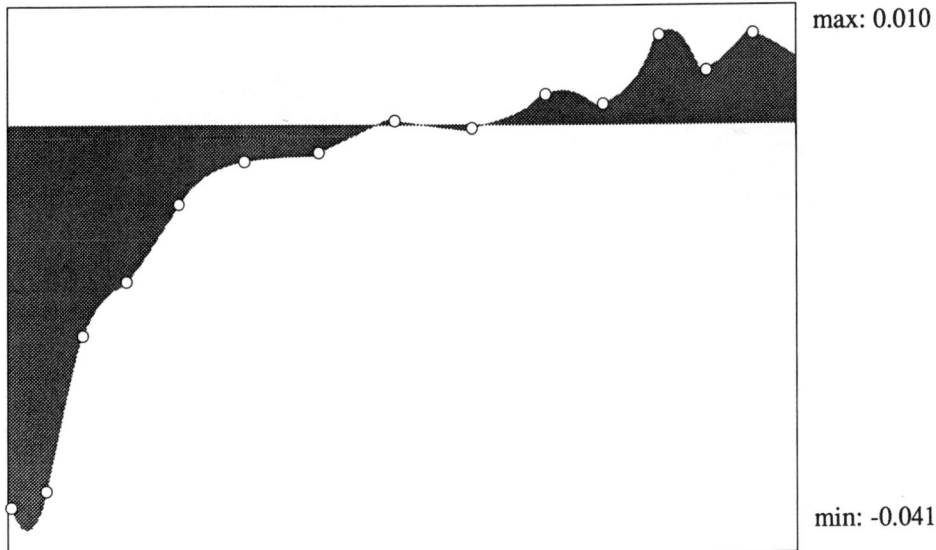

max: 0.010

min: -0.041

FIG. 4.6.   *Original curvature plot.*

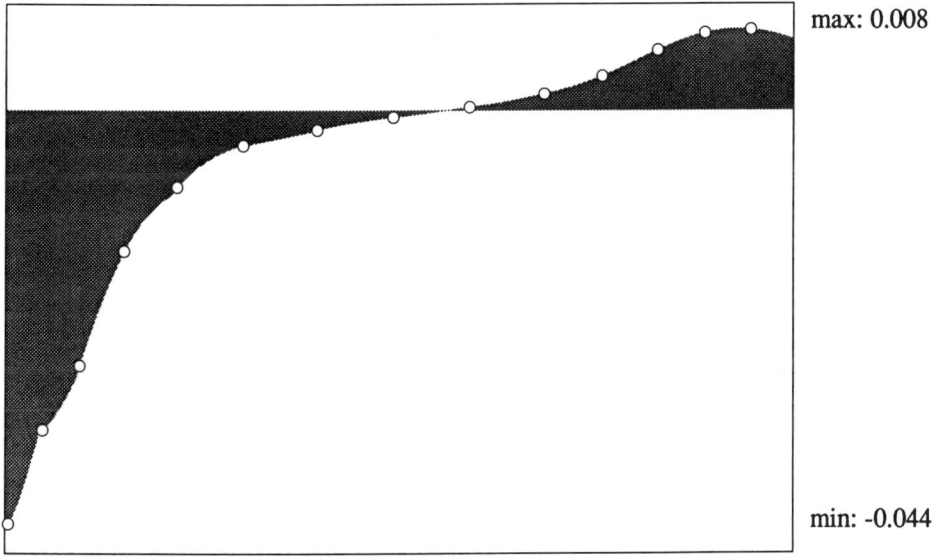

max: 0.008

min: -0.044

FIG. 4.7.   *Curvature plot after two applications of knot removal fairing. Deviation from original polygon: 0.011.*

of them. In the second step, interpret all columns of the resulting control net as B-spline polygons and apply the curve fairing algorithm to each of them. The final control net will correspond to a surface that is fairer than the original one.

We are not attempting here to define the concept of "fairness" for surfaces — this is still an open question (see ([5], [17], [2], [20]). We use an approach

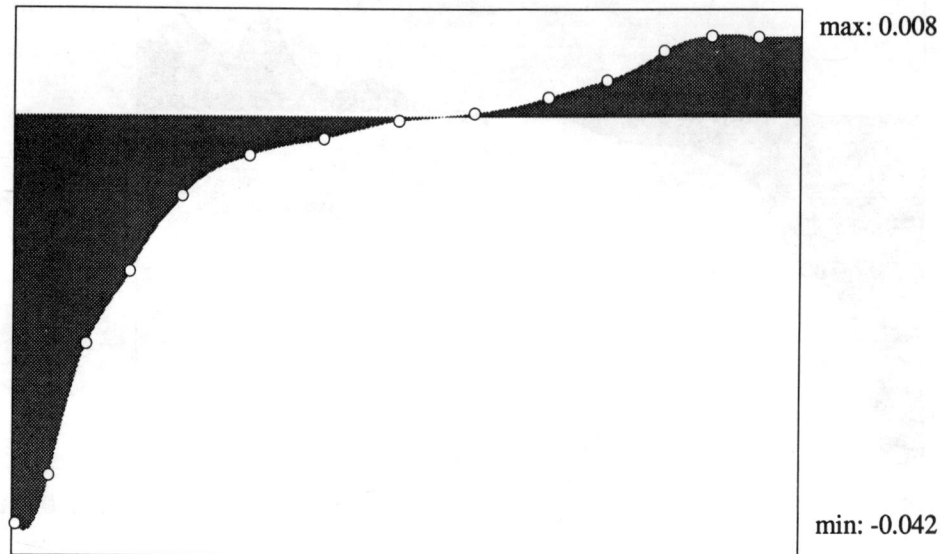

max: 0.008

min: -0.042

FIG. 4.8. *After two applications of degree reduction fairing. Deviation from original polygon:* 0.008.

that was first introduced by Klass [17], a simulation of a method used by stylists to see if the shape of a car is acceptable. A car is placed in a room with parallel fluorescent strip lights on the ceiling. These lights are reflected on the polished car surface, producing so-called "reflection lines." If each of them is fair (see above), then the surface is said to be fair itself. Figure 4.9 gives an example of degree reduction surface fairing. The reflection line that is shown corresponds to a light source perpendicular to the image plane and infinitely high over the surface.

More methods for surface fairing exist; they aim for the enforcement of convexity constraints in tensor product spline surfaces. We mention Andersson et al. [1], Jones [14], and Kaufmann and Klass [15].

4.7. Conclusions

Degree reduction fairing seems to compare favorably to knot removal fairing, in the sense that a curve has to be changed less in order to achieve comparable shape improvements.

Knot removal fairing may be generalized to higher degree curves or to curves with multiple knots; this also holds for degree reduction fairing. We may see this as follows. Any segment of an nth degree B-spline curve is defined by $n + 1$ coefficients, which we may think of as Bézier control vertices. These coefficients depend in a one-to-one manner on $n + 1$ B-spline control vertices. We may now degree reduce the Bézier representation of a polynomial segment, using any of the methods described in [11], [7], [27], and hoping that this results in a shape improvement (again in the sense of removing superfluous information in the curve). We can then recompute the corresponding B-spline

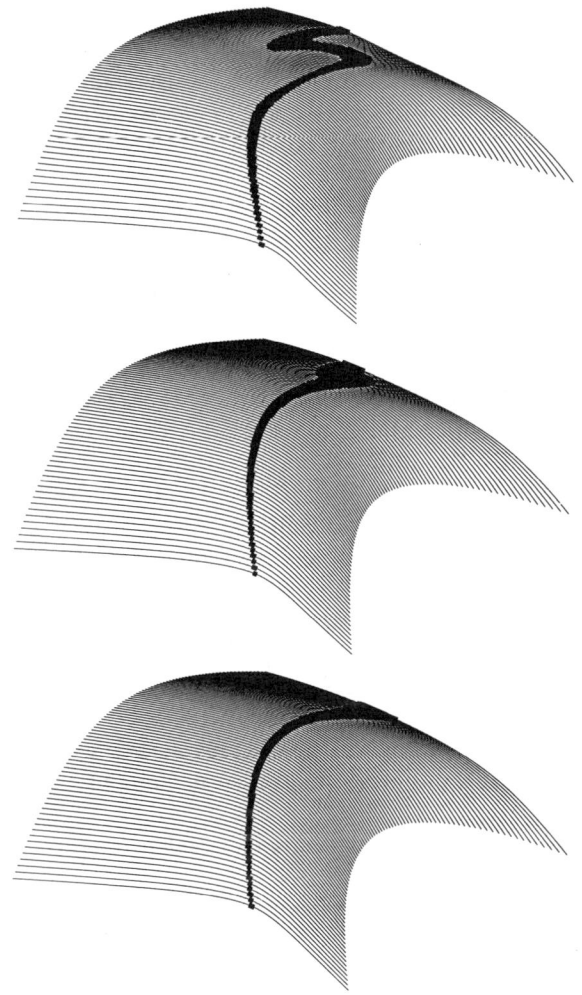

FIG. 4.9. *Surface fairing. Top, a reflection line of the initial surface; middle: after fairing twice; bottom, after fairing four times.*

control vertices. If necessary, they may have to be averaged with already computed control vertices, similar to the use of (4.10).

The tensor product extension performs well. It is an open question how to extend curve fairing methods to more complicated surfaces, for example, those built from triangular interpolants [22] or hybrid schemes [21].

Acknowledgments

This research was supported by National Science Foundation grant DCR-8502858 and Department of Energy contract DE-FG02-87ER25041, both awarded to Arizona State University. I wish to thank N. Sapidis for several fruitful discussions, and a referee for many insightful comments, which improved this article considerably.

References

[1] R. Andersson, E. Andersson, M. Boman, B. Dahlberg, T. Elmroth, and B. Johansson, *The automatic generation of convex surfaces*, in The Mathematics of Surfaces II, R. Martin, ed., Oxford University Press, Oxford, 1987, pp. 427–446.

[2] J. Beck, R. Farouki, and J. Hinds, *Surface analysis methods*, IEEE Comput. Graphics Appl., 6 (1986), pp. 18–36.

[3] G. Birkhoff, *Aesthetic Measure*. Harvard University Press, Cambridge, MA, 1933.

[4] C. de Boor, *A Practical Guide to Splines*. Springer-Verlag, Berlin, New York, 1978.

[5] J. Dill, *An application of color graphics to the display of surface curvature*, Comput. Graphics (Proc. SIGGRAPH), 15 (1981), pp. 153–161.

[6] M. do Carmo, *Differential Geometry of Curves and Surfaces*, Prentice-Hall, Englewood Cliffs, NJ, 1976.

[7] G. Farin, *Curves and Surfaces for Computer Aided Geometric Design*, Academic Press, New York, 1988. Second Edition, 1990.

[8] G. Farin, G. Rein, N. Sapidis, and A. Worsey, *Fairing cubic B-spline curves*, Comput. Aided Geom. Des., 4 (1987), pp. 91–104.

[9] G. Farin and N. Sapidis, *Curvature and the fairness of curves and surfaces*, IEEE Comput. Graphics Appl., 9 (1989), pp. 52–57.

[10] A. Forrest, *Curves and surfaces for computer-aided design*, Ph.D. thesis, University of Cambridge, Cambridge, England, 1968.

[11] ———, *Interactive interpolation and approximation by Bézier polynomials*, Computer J., 15 (1972), pp. 71–79. Reprinted in Comput. Aided Des., 22 (1990), pp. 527-537.

[12] J. Hoschek, *Detecting regions with undesirable curvature*, Comput. Aided Geom. Des., 1 (1984), pp. 183–192.

[13] ———, *Smoothing of curves and surfaces*, Comput. Aided Geom. Des., 2 (1985), pp. 97–105.

[14] A. Jones, *Shape control of curves and surfaces through constrained optimization*, in Geometric Modeling: Algorithms and New Trends, G. Farin, ed., Society for Industrial and Applied Mathematics, Philadelphia, PA, 1987, pp. 265–279.

[15] E. Kaufmann and R. Klass, *Smoothing surfaces using reflection lines for families of splines*, Comput. Aided Des., 20 (1988), pp. 312–316.

[16] J. Kjellander, *Smoothing of bicubic parametric surfaces*, Comput. Aided Des., 15 (1983), pp. 288–293.

[17] R. Klass, *Correction of local surface irregularities using reflection lines*, Comput. Aided Geom. Des., 12 (1980), pp. 73–76.

[18] E. Lee, *Energy, fairness, and a counterexample*, Comput. Aided Des., 22 (1990), pp. 37–40.

[19] K. MacCallum and J.-M. Zhang, *Curve smoothing techniques using B-splines*, The Comput. J, 29 (1986), pp. 564–571.

[20] F. Munchmeyer, *Shape interrogation: a case study*, in Geometric Modeling: Algorithms and New Trends, G. Farin, ed., Society for Industrial and Applied Mathematics, Philadelphia, PA, 1987, pp. 291–301.

[21] J. Peters, *Smooth mesh interpolation with cubic patches*, Comput. Aided Des., 22 (1990), pp. 109–120.

[22] B. Piper, *Visually smooth interpolation with triangular Bézier patches*, in Geometric Modeling: Algorithms and New Trends, G. Farin, ed., Society for Industrial and Applied Mathematics, Philadelphia, PA, 1987, pp. 221–233.

[23] W. Renz, *Interactive smoothing of digitized point data*, Comput. Aided Des., 14 (1982), pp. 267–269.

[24] N. Sapidis, *Algorithms for locally fairing B-spline curves*, Master's thesis, University of Utah, Salt Lake City, UT, 1987.

[25] N. Sapidis and G. Farin, *Automatic fairing algorithm for B-spline curves*, Comput. Aided Des., 22 (1990), pp. 121–129.

[26] B.-Q. Su and D.-Y. Liu, *Computational Geometry*, Academic Press, 1989.

[27] M. Watkins and A. Worsey, *Degree reduction for Bézier curves*, Comput. Aided Des., 20 (1988), pp. 398–405.

General Offset Curves and Surfaces

Eric L. Brechner

5.1. Introduction

Offset surfaces, also known as parallel or canal surfaces, are surfaces that are parallel or "offset" from some given "generator" or "foundation" surface. To find the offset surface, one adds a constant multiple of the foundation surface normal to the foundation surface [10], [11].

Parallel and canal surfaces are mentioned in many classical geometry texts (e.g., [8], [23], [7]), but are usually only of peripheral interest. More recently, these surfaces have been studied in the computer aided geometric design (CAGD) literature under the name "offset surfaces." Much of the work has focused on approximating the often irrational offset surfaces by piecewise polynomial surfaces so that they may be incorporated into current computer aided design (CAD) software ([17], [24], [14], [15], [9], [20], [16]). The properties of offset curves and surfaces have been examined in Papaioannou and Kiritsis [19], Ravani and Ku [22], and Farouki and Neff [10], [11].

An offset surface is sometimes defined as the envelope of circles or spheres centered on a generator curve or surface [5]. In this paper, envelopes of *arbitrary* smooth curves and surfaces centered on generator curves or surfaces are studied. These envelopes are called "general offset" curves and surfaces.

Envelopes, like parallel surfaces, are mentioned in many classical geometry texts (e.g., [8], [23], [18]), but are also usually of peripheral interest. Many of these texts analyze developable surfaces, which are often defined as the envelopes of planes, but the analysis is not extended to envelopes of surfaces other than planes and spheres. More general forms of envelopes have not received wide attention, which led Guggenheimer [13] to state in his text, "The general theory of envelopes is not yet in a completely satisfactory state, but we will not discuss it further except for the case of a family of straight lines where there exists a simple and exhaustive theory" (p. 44).

This paper derives local properties and explicit parameterizations of envelopes formed by smooth "structuring surfaces" originating at every point of a given foundation surface.

FIG. 5.1. *A circle moving along a curve.*

FIG. 5.2. *Envelope of ellipses.*

5.2. Envelopes[1]

Consider a circle moving along a curve (Fig. 5.1). The circle sweeps out a region around the curve. This region might be thought of as the area cut away by the movement of the circle or as a buffer zone around the path of the circle. The boundary of this region forms an envelope.

This type of envelope is often called an offset curve. The analytic and algebraic properties of this type of curve are analyzed by Farouki and Neff [10], [11]. If an ellipse is translated along a curve, then the properties of the resulting envelope are slightly different (Fig. 5.2).

However, the properties of both these envelopes are related. One important relationship derived in Corollary 5.4.1 of this paper is that at a point of contact between one of the moving curves and the envelope the radius of curvature of the envelope is the radius of curvature of the path curve plus or minus the radius of curvature of the moving curve.

---

[1]For a further discussion of envelopes, see Lane [18] or Faux and Pratt [12].

FIG. 5.3. *Circles around a circle.*

Figure 5.3 illustrates this relationship. A small circle moves around a large circle. The radius of the outside envelope is the radius of the large circle plus the radius of the small circle. The radius of the inside envelope is the radius of the large circle minus the radius of the small circle.

Envelopes often appear in applications which involve moving objects. The boundary of the swept region forms an envelope. For a robot arm, the envelope of its movements represents the boundary of the region which should be kept clear. For a mill, the envelope of the movements of the cutting tool represents the part being cut.

The nature of an envelope or boundary of a moving object may be characterized as follows. Every point on the envelope shares a common tangent at that point with the moving object, and for every object position there is at least one point on the object which shares a common tangent with the envelope.

The collection of object positions may be modeled mathematically as a parameterized family of curves or surfaces. The family may be defined implicitly as

(5.1)
$$f(X, u) = 0,$$

where X is a position vector and u is a scalar parameter.

For any fixed value of u, there is an implicitly defined curve or surface in the family. The parameter u might measure distance along the path of an object, while the implicit function (5.1) for a fixed value of u might represent the object.

To compute the envelope, we find points which are on two close members of the same family. Take the two close family members to have the parameters u and $u + \Delta u$. Say X is a point on both family members, i.e., it satisfies both (5.1) and

$$f(X, u + \Delta u) = 0.$$

If we subtract these two equations and divide by Δu, we have

$$\frac{1}{\Delta u}[f(X, u + \Delta u) - f(X, u)] = 0.$$

Now taking the limit as the two family members coalesce, i.e., $\Delta u \to 0$, we find the envelope condition

(5.2) $$\frac{\partial}{\partial u} f(X, u) = 0.$$

(Some care is taken in computing the limit since X depends on Δu.) For multiparameter families, derivatives are taken with respect to each of the parameters and separately set to zero.

The equation for the family, (5.1), and the envelope condition, (5.2), give necessary conditions for a point on the envelope, X, at the parameter value u. However, these conditions are not sufficient to determine the envelope. The envelope must also be tangent to the family of surfaces, as stated earlier.

For the type of envelopes we will be considering, there are three key curves or surfaces. The first is the "structuring surface," which might describe the shape of a cutting tool. The second is the "foundation surface," at whose points the origin of the structuring surface is located. The third is the "envelope," which is formed by locating the origin of the structuring surface at each point on the foundation surface. Envelopes formed in this manner cover the applications discussed earlier.

The discussion of the geometric properties of these envelopes begins with a general derivation of the properties of an $n - 1$ parameter family of surfaces in an n-dimensional Euclidean space and will only consider translations of the surface. These results are then applied to planar curve envelopes ($n = 2$) and further to two-parameter surfaces ($n = 3$), and concrete examples are given in each case. Singularities and other degeneracies are analyzed when encountered.

5.3. Some Notation

For the general case, all surfaces are in an n-dimensional space, parameterized by $n - 1$ parameters, using the following notation.

The n vector, $X = (x_1, x_2, \cdots, x_n)^T$, represents a position in $\mathbf{R}^n$.

$\underline{u}$ represents an $n - 1$ parameter vector $(u_1, u_2, \cdots, u_{n-1})^T$.

Geometric vectors are represented by uppercase letters, parametric vectors by lowercase letters. N-dimensional vectors are represented by characters without underlining, $n - 1$-dimensional vectors by characters with underlining.

$X_{\underline{u}}$ denotes a $\underline{u}$ vector derivative of X.

Vector derivatives are taken in columns, e.g., $X_{\underline{u}}$ has n rows representing each position coordinate and $n-1$ columns representing each parameter derivative.

N is used to represent unit normal vectors.

$Xf(\underline{u})$ represents the foundation surface. (Xf is a function with values in n-space.)

$Xf_{\underline{u}}(\underline{u})$ represents a spanning set for the tangent space of the foundation surface. ($Xf_{\underline{u}}(\underline{u})$ represents the n by $n-1$ matrix of partial derivatives of Xf. Each column represents a vector in the tangent plane.)

$Nf(\underline{u})$ represents the unit normal of the foundation surface.

$Xe(\underline{u})$ represents the envelope surface.

$s(X) = 0$ is an implicit definition of the structuring surface. ($s(X)$ is real valued.)

$Xs(\underline{t})$ represents a parameterization of the structuring surface, where $\underline{t}$ represents an $n-1$ parameter vector $(t_1, t_2, \cdots, t_{n-1})^T$.

5.4. The Geometry of General Offsets

General offsets are envelopes formed by translations of the origin of a structuring surface. In the given notation such envelopes satisfy

$$(5.3a) \qquad\qquad s(X - Xf(\underline{u})) = 0,$$

where $X - Xf(\underline{u})$ provides translations parameterized by $\underline{u}$ of the origin of the structuring surface $s(X) = 0$. This representation covers offset surfaces and surfaces milled by cutting tools of fixed orientation. The envelope condition is

$$\frac{\partial}{\partial \underline{u}} s(X - Xf(\underline{u})) = 0,$$

or

$$(5.3b) \qquad\qquad s_x X f_{\underline{u}} = 0,$$

where s_x is the $1 \times n$ gradient of $s(X)$, and the product represents matrix multiplication.

Under the conditions defined below, the envelope may be thought of as being parameterized by the same parameters as the foundation surface, i.e., $Xe(\underline{u})$. If $X (= Xe(\underline{u}))$ is a point on the envelope then (5.3a) is satisfied, which means $Xe(\underline{u}) - Xf(\underline{u})$ represents a point on the structuring surface. Recall that $\underline{t}$ is used to parameterize the structuring surface $Xs(\underline{t})$. Thus, there must be some $\underline{t} = \underline{t}(\underline{u})$ such that $Xe(\underline{u}) - Xf(\underline{u}) = Xs(\underline{t}(\underline{u}))$, or

$$(5.4) \qquad\qquad Xe(\underline{u}) = Xf(\underline{u}) + Xs(\underline{t}(\underline{u})).$$

Any function $\underline{t}(\underline{u})$ will produce values for $Xe(\underline{u})$ that satisfy (5.3a). However, to compute points on the envelope, $\underline{t}(\underline{u})$ must also satisfy (5.3b), i.e.,

$$s_x(Xe(\underline{u}) - Xf(\underline{u}))Xf_{\underline{u}} = s_x(Xs(\underline{t}(\underline{u})))Xf_{\underline{u}} = 0,$$

where $s_x(Xs(\underline{t}(\underline{u})))$ is the value of the gradient of $s(X)$ evaluated at $Xs(\underline{t}(\underline{u}))$.

To find the function $\underline{t}(\underline{u})$ and thus the envelope $Xe(\underline{u})$, we need to know more about the gradient s_x.

5.4.1. Implicit Representation of a Surface.

The following known result shows that without loss of generality it can be assumed that the gradient of $s(X)$ is nonzero near a point X on the structuring surface and that at the point X the gradient is the surface normal vector.

THEOREM 5.4.1. (Implicit representation of a surface.) *Given some fixed value of $\underline{t} = \underline{t}_0$ at which the structuring surface $Xs(\underline{t})$ has continuous derivatives with respect to $\underline{t}$ and an $n-1$-dimensional tangent space, there exists a continuous, real-valued function $s(X)$ with continuous derivatives $s_x(X)$ for which $Xs(\underline{t})$ satisfies $s(Xs(\underline{t})) \equiv 0$ near $\underline{t} = \underline{t}_0$ and $x_x(X)$ represents the normal of the surface $s(X) = 0$ at $X = Xs(\underline{t}_0)$.*

Proof. If an $n-1$-dimensional tangent space exists for the structuring surface at $\underline{t} = \underline{t}_0$, then the tangent vectors of the structuring surface are linearly independent at $\underline{t} = \underline{t}_0$. Therefore, the column vectors of $Xs_{\underline{t}}(\underline{t}_0)$ taken with the normal vector $Ns(\underline{t}_0)$ form a basis for n space.

Using this information, we define a new coordinate system where the first $n-1$ basis vectors are given by the tangents to the structuring surface and the last basis vector is given by the normal to the structuring surface. We denote the coordinates in the tangent directions by the $n-1$ coordinate vector $\underline{Z}$ and the coordinate in the normal direction by the scalar y.

The vector $(\underline{Z}, y)^T$ may be written in terms of an arbitrary vector X in the original coordinate system as follows:

$$\underline{Z} = [Xs_{\underline{t}}(\underline{t}_0)^T Xs_{\underline{t}}(\underline{t}_0)]^{-1} Xs_{\underline{t}}(\underline{t}_0)^T (X - Xs(\underline{t}_0)),$$
$$y = Ns(\underline{t}_0)^T (X - Xs(\underline{t}_0)).$$

We propose to find a function $y = y(\underline{Z})$ that gives a local representation of the structuring surface.

In the $(\underline{Z}, y)^T$ coordinate system, the parametric equation for the surface becomes

(5.5)
$$\underline{Z}s(\underline{t}) = [Xs_{\underline{t}}(\underline{t}_0)^T Xs_{\underline{t}}(\underline{t}_0)]^{-1} Xs_{\underline{t}}(\underline{t}_0)^T (Xs(\underline{t}) - Xs(\underline{t}_0)),$$
$$ys(\underline{t}) = Ns(\underline{t}_0)^T (Xs(\underline{t}) - Xs(\underline{t}_0)),$$

and its tangents are given as

$$\underline{Z}s_{\underline{t}}(\underline{t}) = [Xs_{\underline{t}}(\underline{t}_0)^T Xs_{\underline{t}}(\underline{t}_0)]^{-1} Xs_{\underline{t}}(\underline{t}_0)^T Xs_{\underline{t}}(\underline{t}),$$
$$ys_{\underline{t}}(\underline{t}) = Ns(\underline{t}_0)^T Xs_{\underline{t}}(\underline{t}).$$

Evaluating the tangents at $\underline{t}_0$, we find

(5.6) $\underline{Z}s_{\underline{t}}(\underline{t}_0) = \underline{I}$ and $ys_{\underline{t}}(\underline{t}_0) = 0$ (by inspection).

We consider the implicit $n - 1$ system of equations for $\underline{t}$,

$$\underline{Z} = \underline{Z}s(\underline{t}).$$

We know $\underline{Z}s$ inherits continuous derivatives from Xs, and we know from (5.6) that its derivatives are linearly independent at $\underline{t}_0$. We also know from (5.5) that when $\underline{t} = \underline{t}_0$, $\underline{Z} = 0$. Therefore, applying the Implicit Function Theorem,[2] we find that locally $\underline{t}$ can be written in terms of $\underline{Z}$, i.e.,

$$\underline{t} = \underline{t}(\underline{Z}) \quad \text{for } \underline{Z} \text{ near } \underline{Z} = 0.$$

Now, define $s(X)$ as follows:

$$s(X) \equiv y(\underline{t}(\underline{Z})) = Ns(\underline{t}_0)^T[X - Xs(\underline{t}(\underline{Z}))],$$

where $\underline{Z}$ is computed from X as shown earlier. By definition, Xs is locally a solution to $s(X) = 0$. Calculating s_x, we find

$$s_x(X) = Ns(\underline{t}_0)^T[\underline{I} - Xs_{\underline{t}}\underline{t}_{\underline{z}}\underline{Z}_x],$$

where

$$\underline{Z}_x = [Xs_{\underline{t}}(\underline{t}_0)^T Xs_{\underline{t}}(\underline{t}_0)]^{-1} Xs_{\underline{t}}(\underline{t}_0)^T.$$

Evaluated at $Xs(\underline{t}_0)$, we get

$$s_x(Xs(\underline{t}_0)) = Ns(\underline{t}_0)^T[\underline{I} - Xs_{\underline{t}}(\underline{t}_0)\underline{t}_{\underline{z}}\underline{Z}_x].$$

At $\underline{t} = \underline{t}_0$ we have that $\underline{Z} = 0$, $\underline{t}_{\underline{z}} = \underline{I}$, and $\underline{Z}_x$ is as before. Since $Ns(\underline{t}_0)^T Xs_{\underline{t}}(\underline{t}_0) = 0$,

$$s_x(Xs(\underline{t}_0)) = Ns(\underline{t}_0)^T. \qquad \square$$

5.4.2. Collinearity of Surface Normals. Now a relationship, $\underline{t}(\underline{u})$, between the parameterizations $\underline{t}$ and $\underline{u}$ may be derived. First, we use the previous result to define a geometric condition for points on the envelope.

THEOREM 5.4.2. (Collinearity of surface normals.) *Given some fixed value of $\underline{u}$ at which the function describing the foundation surface has an $n - 1$-dimensional tangent space, consider all the points X which satisfy*

$$s(X - Xf(\underline{u})) = 0.$$

---

[2]The Implicit Function Theorem states that a nonlinear system of n equations, $\underline{f}(\underline{x}, y) = 0$, with parameters, $\underline{x}$, and n unknowns, y, has a unique differentiable solution, $y(\underline{x})$, in the neighborhood of a given solution, $\underline{f}(\underline{x}_0, y_0) = 0$, if $\underline{f}_{\underline{x}}$ and $\underline{f}_{\underline{y}}$ are continuous and $\underline{f}_{\underline{y}}$ forms a nonsingular matrix at $(\underline{x}_0, y_0)$. A precise statement of the theorem may be found in most Real Analysis texts (e.g., Protter and Morrey [21]).

In the case above, the parameters are $\underline{Z}$, the unknowns are $\underline{t}$, and the equations are $\underline{Z}s(\underline{t}) - \underline{Z} = 0$. At $\underline{t} = \underline{t}_0$, $\underline{Z} = 0$ the equations are solved, the derivatives with respect to $\underline{Z}$ and $\underline{t}$ are continuous, and $\underline{Z}s_{\underline{t}}(\underline{t}_0) = \underline{I}$, which is clearly nonsingular. Therefore, there exists a unique differentiable function, $\underline{t}(\underline{Z})$, which solves the equations near $\underline{t} = \underline{t}_0$, $\underline{Z} = 0$.

If the structuring surface at one of the points, $X - Xf(\underline{u})$, has an $n - 1$-dimensional tangent space and an envelope exists then the point X is on the envelope only if the normals of the foundation and structuring surfaces, evaluated at $\underline{u}$ and $X - Xf(\underline{u})$, respectively, are collinear.

Furthermore, if the envelope exists and has an $n - 1$-dimensional tangent space at the point X, then its normal at X is collinear with the normals of the foundation and structuring surfaces, evaluated at $\underline{u}$ and $X - Xf(\underline{u})$, respectively.

Proof. If X is a point on the envelope then both (5.3a) and (5.3b) are satisfied at that point. Equation (5.3b),

$$s_x X F_{\underline{u}} = 0,$$

states that the gradient of $s(X)$ is orthogonal to the foundation surface tangent plane. Therefore, we have that the gradient of $s(X)$ is the normal of the foundation surface. The gradient of $s(X)$ is also the normal of the structuring surface. Thus, we have

$$\frac{s_x^T}{|s_x|} = \pm N f(\underline{u}).$$

If the envelope exists, then from (5.3a)

$$s(Xe(\underline{u}) - Xf(\underline{u})) = 0.$$

Taking the derivative with respect to $\underline{u}$,

$$\frac{\partial}{\partial \underline{u}} s(Xe(\underline{u}) - Xf(\underline{u})) = 0,$$
$$s_x(Xe_{\underline{u}} - Xf_{\underline{u}}) = 0,$$

and, using (5.3b),

$$s_x Xe_{\underline{u}} = 0.$$

Thus, if the envelope has an $n - 1$-dimensional tangent space, then the gradient of $s(X)$ is normal to the envelope, i.e.,

$$\frac{s_x^T}{|s_x|} = \pm N f(\underline{u}) = \pm N e(\underline{u}). \qquad \square$$

So for envelopes formed by translating a fixed surface, there is an interesting geometric result. The normals of the structuring surface, the foundation surface, and the envelope surface are all collinear for any given parameter value $\underline{u}$.

5.4.3. Existence of the Envelope.   At this point there is enough information to determine the envelope. From Theorem 5.4.2 the relationship between $\underline{u}$ and $\underline{t}$ in (5.4) is given as

(5.7) $$N s(\underline{t}) = \pm N f(\underline{u}),$$

which is equivalent to the statement that the normal of the structuring surface
is orthogonal to all the tangent vectors of the foundation surface, i.e.,

$$Ns^T(\underline{t})Xf_{\underline{u}}(\underline{u}) = 0.$$

The equation $Ns(\underline{t}) = \pm Nf(\underline{u})$ forms a redundant system of n equations.
The system is redundant because Ns and Nf are constrained to be unit
vectors, i.e., if $n-1$ components of Ns or Nf are determined, then the value
of the last component may be calculated from the other components except
for its sign. Thus, there are $n-1$ equations for $n-1$ unknowns $\underline{t}$ in terms of
$n-1$ parameters $\underline{u}$. These unknowns can be found locally, with well-known
constraints, using the Implicit Function Theorem.

THEOREM 5.4.3. (Existence of the envelope.) *Assume that at particular
values of $\underline{u}$ and $\underline{t}$, $Nf_{\underline{u}}(\underline{u})$ and $Ns_{\underline{t}}(\underline{t})$ are continuous and the columns of $Ns_{\underline{t}}(\underline{t})$
form a linearly independent set of vectors. If*

$$Ns(\underline{t}) = \pm Nf(\underline{u}),$$

then there exists, locally, a function $\underline{t} = \underline{t}(\underline{u})$, as well as a surface defined by

$$Xe(\underline{u}) = Xf(\underline{u}) + Xs(\underline{t}(\underline{u}))$$

with continuous derivatives that locally form an envelope.

Proof. Let the particular values of $\underline{u}$ and $\underline{t}$ be $\underline{u}_0$ and $\underline{t}_0$ respectively. Let
$\underline{\underline{H}}$ be the $n \times n-1$ matrix whose columns form an orthonormal basis for the
tangent space of $Xf(\underline{u})$ at $\underline{u} = \underline{u}_0$ (as derived from the columns of $Xf_{\underline{u}}(\underline{u}_0)$).

The equation

$$\underline{\underline{H}}^T Ns(\underline{t}) = \pm \underline{\underline{H}}^T Nf(\underline{u})$$

is satisfied at $\underline{u} = \underline{u}_0$ and $\underline{t} = \underline{t}_0$. For this system of $n-1$ equations, with
an arbitrary value of $\underline{u}$, we determine $n-1$ components of $Ns(\underline{t})$. (Here
$Ns(\underline{t})$ is thought of as a function of $Nf(\underline{u})$.) The last component of $Ns(\underline{t})$
has the value $Nf(\underline{u}_0)^T Ns(\underline{t})$. Its magnitude may be computed from the other
$n-1$ components of $Ns(\underline{t})$. Its sign is the plus or minus sign from the original
matching of normals, $Ns(\underline{t}_0) = \pm Nf(\underline{u}_0)$, and the sign will not change without
$Nf(\underline{u}_0)^T Ns(\underline{t})$ vanishing.

The magnitude of $Nf(\underline{u}_0)^T Ns(\underline{t})$ vanishes only when the magnitude of the
same component of $Nf(\underline{u})$ vanishes, i.e., when $Nf(\underline{u}_0)^T Nf(\underline{u}) = 0$. There, we
can determine the sign of $Nf(\underline{u}_0)^T Ns(\underline{t})$ for values of $\underline{u}$ in the region about
$\underline{u}_0$ where $Nf(\underline{u})$ does not rotate away from $Nf(\underline{u}_0)$ by 90 degrees or more.
Outside that region we might need to redefine $\underline{u}_0$ and $\underline{t}_0$ and start over.

To apply the Implicit Function Theorem to the system of $n-1$ equations

$$\underline{\underline{H}}^T Ns(\underline{t}) = \pm \underline{\underline{H}}^T Nf(\underline{u})$$

and solve for $\underline{t}$ in terms of $\underline{u}$ near $\underline{u}_0$ and $\underline{t}_0$, we need to satisfy a few conditions.

The first condition is that the derivative of the right-hand side with respect to $\underline{u}$, $\pm \underline{\underline{H}}^T N f_{\underline{u}}(\underline{u})$, is continuous. This condition is satisfied by the assumptions of the theorem.

The second condition is that the derivative of the left-hand side with respect to $\underline{t}$, $\underline{\underline{H}}^T N s_{\underline{t}}(\underline{t})$, is continuous. This condition is also met by the assumptions of the theorem.

The last condition, because we are solving for $\underline{t}$, is that the derivative of the left-hand side with respect to $\underline{t}$, $\underline{\underline{H}}^T N s_{\underline{t}}(\underline{t})$, forms a nonsingular matrix. This condition is met because the columns of both $\underline{\underline{H}}$ and $N s_{\underline{t}}(\underline{t})$ are given to be linearly independent, so the product $\underline{\underline{H}}^T N s_{\underline{t}}(\underline{t})$ forms a nonsingular matrix.

Thus, we may apply the Implicit Function Theorem, which states there exists a unique, continuously differentiable function

$$\underline{t} = \underline{t}(\underline{u})$$

such that near $\underline{u}_0$ and $\underline{t}_0$

$$\underline{\underline{H}}^T N s(\underline{t}(\underline{u})) = \pm \underline{\underline{H}}^T N f(\underline{u}).$$

If we further restrict the region in which $\underline{t}(\underline{u})$ is valid to include only values of $\underline{u}$ inside the area about $\underline{u}_0$ where $N f(\underline{u}_0)^T N f(\underline{u}) > 0$, then in that local region

$$N f(\underline{u}) = \pm N s(\underline{t}(\underline{u})).$$

The derivative, also known as the Jacobian, of the function $\underline{t}(\underline{u})$ may be calculated from

$$N f_{\underline{u}} = \pm N s_{\underline{t}} \underline{t}_{\underline{u}}.$$

Likewise, the tangent plane of the envelope may be found by manipulating (5.4) as follows:

(5.8)
$$X e(\underline{u}) = X f(\underline{u}) + X s(\underline{t}(\underline{u})),$$
$$X e_{\underline{u}} = X f_{\underline{u}} + X s_{\underline{t}} \underline{t}_{\underline{u}}.$$

We know from Theorem 5.4.2 that the normal to the tangent plane of the envelope for some value $\underline{u}$ is orthogonal to the tangent plane of its corresponding structuring surface, $X s_{\underline{t}}(\underline{t}(\underline{u}))$, and, therefore, $X e(\underline{u})$ locally forms an envelope of the family of translated structuring surfaces. □

Note that if we choose to parameterize the envelope by $\underline{t}$ instead of $\underline{u}$ then all the requirements for the envelope would be the same except that $N f_{\underline{u}}(\underline{u})$, rather than $N s_{\underline{t}}(\underline{t})$, would be required to form a linearly independent set of vectors. Therefore, we really only require that one or the other set of derivatives of the normals be linearly independent in order to use this local representation of the envelope.

5.4.4. Geometry of the Envelope.

Now we may compute the geometric properties of the generated envelope. Consider an arbitrarily parameterized surface $X(\underline{u})$ and the relationship between its tangents and the derivatives of its normal vector. The normal is a unit vector. Therefore,

$$N^T N = 1 \Rightarrow N^T N_{\underline{u}} = 0.$$

Thus, the derivatives of the normal vector lie in the tangent plane. This means it is possible to write the derivatives of the normal vector as linear combinations of the tangent vectors, i.e., there is a linear map between the tangent vectors and the normal derivatives.

This is known as the differential Gauss map [7], which may be written as

(5.9) $$N_{\underline{u}}(\underline{u}) = -X_{\underline{u}}(\underline{u})\,\underline{\kappa}\,(\underline{u}),$$

where $\underline{\kappa}(\underline{u})$ is an $n - 1 \times n - 1$ matrix. The differential Gauss map is, among other things, self adjoint. The eigenvalues of the map represent the principle curvatures of the surface at $X(\underline{u})$; the eigenvectors of the map are orthogonal and represent the principle curvature directions. The Gaussian curvature may be calculated by taking the determinant of $\underline{\kappa}$; the mean curvature may be calculated by dividing the trace of $\underline{\kappa}$ by $n - 1$.

The problem now becomes finding $\underline{\kappa}$ for the envelope. To do this, we use a second matrix $\underline{M}$ which relates the tangent vectors of the foundation surface to the tangent vectors of the structuring surface as follows:

(5.10) $$X f_{\underline{u}} = X s_{\underline{t}}\,\underline{M}.$$

This relation is exact when the normals of the structuring and foundation surfaces are collinear, which corresponds to the envelope condition. Note that $\underline{M}$ is nonsingular when both $X f_{\underline{u}}$ and $X s_{\underline{t}}$ have linearly independent column vectors.

THEOREM 5.4.4. (Geometry of the envelope.) *Assume that for particular values of $\underline{u}$ and $\underline{t}$, $X f_{\underline{u}}(\underline{u})$, $X s_{\underline{t}}(\underline{t})$, $N f_{\underline{u}}(\underline{u})$, and $N s_{\underline{t}}(\underline{t})$ are continuous. Define the relationships between these quantities as*

$$X f_{\underline{u}} = X s_{\underline{t}}\,\underline{M}, \quad N f_{\underline{u}} = -X f_{\underline{u}}\,\underline{\kappa f}, \quad \text{and} \quad N s_{\underline{t}} = -X s_{\underline{t}}\,\underline{\kappa s}.$$

Assume further that $\underline{\kappa s}(\underline{t})$ is nonsingular and

$$\sigma = \mathrm{signum}(\det[\underline{I} \pm \underline{M}^{-1}\underline{\kappa s}^{-1}\underline{M}\,\underline{\kappa f}])$$

is nonzero. If

$$N s(\underline{t}) = \pm N f(\underline{u})$$

then there exists, locally, a function $\underline{t} = \underline{t}(\underline{u})$, as well as an envelope parameterized by $\underline{u}$ whose differential Gauss map is

(5.11) $$\underline{\kappa e} = \sigma[\underline{I} \pm \underline{M}^{-1}\underline{\kappa s}^{-1}\underline{M}\,\underline{\kappa f}]^{-1}\underline{\kappa f}.$$

Furthermore, if $\underline{\kappa f}$ is nonsingular, then

(5.12) $$\sigma\underline{\kappa e}^{-1} = \underline{\kappa f}^{-1} \pm \underline{t_{\underline{u}}}^{-1}\underline{\kappa s}^{-1}\underline{t_{\underline{u}}}.$$

Proof. The columns of $Ns_{\underline{t}}$ form a linear independent set of vectors because the columns of $Xs_{\underline{t}}$ are linearly independent and the mapping $\underline{\underline{\kappa s}}$ between them is nonsingular. Therefore, a continuously differentiable envelope locally exists according to Theorem 5.4.3.

Starting with (5.8) and performing a number of substitutions, we find

(5.13)

$$Xe_{\underline{u}} = Xf_{\underline{u}} + Xs_{\underline{t}}\underline{t}_{\underline{u}},$$
$$Xe_{\underline{u}} = Xf_{\underline{u}} + Xf_{\underline{u}}\underline{\underline{M}}^{-1}\underline{t}_{\underline{u}},$$
$$Xe_{\underline{u}} = Xf_{\underline{u}}[\underline{\underline{I}} + \underline{\underline{M}}^{-1}\underline{t}_{\underline{u}}],$$
$$Xe_{\underline{u}}[\underline{\underline{I}} + \underline{\underline{M}}^{-1}\underline{t}_{\underline{u}}]^{-1} = Xf_{\underline{u}},$$
$$-Xe_{\underline{u}}[\underline{\underline{I}} + \underline{\underline{M}}^{-1}\underline{t}_{\underline{u}}]^{-1}\underline{\kappa f} = Nf_{\underline{u}},$$
$$-Xe_{\underline{u}}[\underline{\underline{I}} + \underline{\underline{M}}^{-1}\underline{t}_{\underline{u}}]^{-1}\underline{\kappa f} = Nf_{\underline{u}}.$$

At this point, we may replace $Nf_{\underline{u}}$ with $Ne_{\underline{u}}$, give or take a sign change, because the normals of the two surfaces are collinear. The sign may be computed from (5.13), which defines the map between the tangents of the foundation surface and the tangents of the envelope at $\underline{u}$. The direction of the envelope normal will depend on the determinant of $[\underline{\underline{I}} + \underline{\underline{M}}^{-1}\underline{t}_{\underline{u}}]$.

To better understand the matrix $[\underline{\underline{I}} + \underline{\underline{M}}^{-1}\underline{t}_{\underline{u}}]$, we manipulate (5.10):

$$Xf_{\underline{u}} = Xs_{\underline{t}}\,\underline{\underline{M}},$$
$$Nf_{\underline{u}} = -Xs_{\underline{t}}\,\underline{\underline{M}}\,\underline{\kappa f}.$$

From Theorem 5.4.3, $Nf_{\underline{u}} = \pm Ns_{\underline{t}}\underline{t}_{\underline{u}}$, which implies

(5.14)

$$\pm Ns_{\underline{t}}\underline{t}_{\underline{u}} = -Xs_{\underline{t}}\,\underline{\underline{M}}\,\underline{\kappa f},$$
$$\pm Xs_{\underline{t}}\,\underline{\underline{\kappa s}}\,\underline{t}_{\underline{u}} = Xs_{\underline{t}}\,\underline{\underline{M}}\,\underline{\kappa f},$$
$$\pm\underline{\underline{\kappa s}}\,\underline{t}_{\underline{u}} = \underline{\underline{M}}\,\underline{\kappa f}.$$

The last step is possible because the column vectors of $Xs_{\underline{t}}$ are linearly independent.

Now, using the face that $\underline{\underline{\kappa s}}$ is nonsingular,

$$\underline{\underline{I}} + \underline{\underline{M}}^{-1}\underline{t}_{\underline{u}} = \underline{\underline{I}} \pm \underline{\underline{M}}^{-1}\underline{\underline{\kappa s}}^{-1}\underline{\underline{M}}\,\underline{\kappa f},$$

and we can rewrite (5.13) as

$$Xe_{\underline{u}} = Xf_{\underline{u}}[\underline{\underline{I}} \pm \underline{\underline{M}}^{-1}\underline{\underline{\kappa s}}^{-1}\underline{\underline{M}}\,\underline{\kappa f}].$$

Thus,

$$-Xe_{\underline{u}}[\underline{\underline{I}} \pm \underline{\underline{M}}^{-1}\underline{\underline{\kappa s}}^{-1}\underline{\underline{M}}\,\underline{\kappa f}]^{-1}\underline{\kappa f} = \sigma Ne_{\underline{u}},$$

which implies that the differential Gauss map for the envelope is given as

$$\underline{\underline{\kappa e}} = \sigma[\underline{\underline{I}} \pm \underline{\underline{M}}^{-1}\underline{\underline{\kappa s}}^{-1}\underline{\underline{M}}\,\underline{\kappa f}]^{-1}\underline{\kappa f}.$$

Furthermore, if $\underline{\underline{\kappa f}}$ is nonsingular, then using the above and (5.14) we find

$$\sigma \underline{\underline{\kappa e}}^{-1} = \underline{\underline{\kappa f}}^{-1} \pm \underline{\underline{t_u}}^{-1} \underline{\underline{\kappa s}}^{-1} \underline{\underline{t_u}}. \qquad \square$$

COROLLARY 5.4.1. (General offset curves.) *Assume that for particular values of u and t,* $Xf'(u) = (xf'(u), yf'(u))^T$ *and* $Xs'(t) = (xs'(t), ys'(t))^T$ *are continuous and have nonzero norms. Let*

$$Nf(u) = \frac{1}{|Xf'(u)|} \begin{pmatrix} -yf'(u) \\ xf'(u) \end{pmatrix} \quad \text{and} \quad Ns(t) = \frac{1}{|Xs'(t)|} \begin{pmatrix} -ys'(t) \\ xs'(t) \end{pmatrix}.$$

Define the curvature of these curves implicitly by

$$Nf' = -Xf' \kappa f \quad \text{and} \quad Ns' = -Xs' \kappa s.[3]$$

Assume further that $\kappa s(t)$ *is nonzero and*

$$\sigma = \text{signum} \left(1 \pm \frac{\kappa f}{\kappa s} \right)$$

is nonzero. If

$$Ns(t) = \pm Nf(u),$$

then there exists, locally, a function $t = t(u)$, *as well as an envelope parameterized by u whose curvature is*

$$\kappa e = \sigma \frac{\kappa f}{1 \pm (\kappa f / \kappa s)} = \frac{\kappa f}{|1 \pm (\kappa f / \kappa s)|}.$$

Furthermore, if κf *is nonzero, then*

$$\sigma \rho e = \rho f \pm \rho s,$$

where ρe *is the radius of curvature of the envelope curve.*

COROLLARY 5.4.2. (General offset surfaces.) *Assume that for particular values of* $\underline{u} = (u, v)$ *and* $\underline{t} = (t, w)$,

$$Xf_{\underline{u}}(\underline{u}) = \begin{pmatrix} xf_u(u,v) & xf_v(u,v) \\ yf_u(u,v) & yf_v(u,v) \\ zf_u(u,v) & zf_v(u,v) \end{pmatrix} \quad \text{and} \quad Xs_{\underline{t}}(\underline{t}) = \begin{pmatrix} xs_t(t,w) & xs_w(t,w) \\ ys_t(t,w) & ys_w(t,w) \\ zs_t(t,w) & zs_w(t,w) \end{pmatrix}$$

are continuous. Assume further that

$$Nf(\underline{u}) \equiv \frac{Xf_u(u,v) \wedge Xf_v(u,v)}{|Xf_u(u,v) \wedge Xf_v(u,v)|} \quad \text{and} \quad Ns(\underline{t}) \equiv \frac{Xs_t(t,w) \wedge Xs_w(t,w)}{|Xs_t(t,w) \wedge Xs_w(t,w)|}$$

---

[3] The differential Gauss map in two dimensions is another form of the planar version of the second Frenet equation [23],

$$N' = -t\kappa \Rightarrow \frac{N'}{|X'|} = -\frac{X'}{|X'|}\kappa \Rightarrow N' = -X'\kappa.$$

So κ is the curvature, and κ^{-1} is the radius of curvature, ρ.

have continuous derivatives. Define the relationships between these sets of vectors as follows:

$$X f_{\underline{u}} = X s_{\underline{t}} \, \underline{\underline{M}}, \quad N f_{\underline{u}} = -X f_{\underline{u}} \, \underline{\underline{\kappa f}}, \quad \text{and} \quad N s_{\underline{t}} = -X s_{\underline{t}} \, \underline{\underline{\kappa s}}.$$

Assume that $\underline{\underline{\kappa s}}(\underline{t})$ is nonsingular and

$$\sigma = \text{signum}(\det[\underline{\underline{I}} \pm \underline{\underline{M}}^{-1} \underline{\underline{\kappa s}}^{-1} \underline{\underline{M}} \, \underline{\underline{\kappa f}}])$$

is nonzero. If

$$N s(\underline{t}) = \pm N f(\underline{u}),$$

then there exists, locally, a function $\underline{t} = \underline{t}(\underline{u})$, as well as an envelope parameterized by $\underline{u}$ whose differential Gauss map is

$$\underline{\underline{\kappa e}} = \sigma[\underline{\underline{I}} \pm \underline{\underline{M}}^{-1} \underline{\underline{\kappa s}}^{-1} \underline{\underline{M}} \, \underline{\underline{\kappa f}}]^{-1} \underline{\underline{\kappa f}}.$$

Furthermore, if $\underline{\underline{\kappa f}}$ is nonsingular, then

$$\sigma \underline{\underline{\kappa e}}^{-1} = \underline{\underline{\kappa f}}^{-1} \pm \underline{t}_{\underline{u}}^{-1} \underline{\underline{\kappa s}}^{-1} \underline{t}_{\underline{u}}.$$

5.4.5. Discussion of Results.

The equation of $\sigma \underline{\underline{\kappa e}}^{-1} = \underline{\underline{\kappa f}}^{-1} \pm \underline{t}_{\underline{u}}^{-1} \underline{\underline{\kappa s}}^{-1} \underline{t}_{\underline{u}}$ states that the inverse curvature matrices of the three surfaces are additive. The expression $\underline{t}_{\underline{u}}^{-1} \underline{\underline{\kappa s}}^{-1} \underline{t}_{\underline{u}}$ is the inverse curvature matrix for the structuring surface reparameterized as a function of $\underline{u}$. The inverse curvature matrices have the same eigenvectors as the original matrices, but the reciprocal eigenvalues. The eigenvalues of the inverse curvature matrices are the principle radii of curvature.

The equation for the envelope, (5.4), is also additive. Therefore, a certain reciprocity is introduced between the envelope and the foundation surface. The foundation surface required to create a certain envelope may be calculated by using the envelope itself as a foundation surface. The resulting envelope formed in this second case will be the original foundation surface. This is assuming the same structuring surface is used in both cases, being careful to reflect it through its origin in the second case to have the correct sign on the normal and curvature.

If $\underline{\underline{\kappa f}}$ is singular, implying that the foundation surface is flat in one or more directions, then (5.11) indicates that not only will $\underline{\underline{\kappa e}}$ be singular, but the envelope will be flat in exactly the same parameter directions as the foundation surface.

The curvature does not depend on the distance from the origin of the structuring surface to the point on the envelope. This means that the position of the origin is only important for determining the position of the envelope, not its shape.

The key to finding the envelope lies in solving (5.7), i.e., finding a point on the structuring surface whose normal matches a given normal on the foundation surface. This involves solving a nonlinear system of $n-1$ equations for $n-1$ unknowns. However, for a number of applications, including three-axis milling, the structuring surface is convex in the area of interest and, therefore, may be parameterized by its normal except along flat sections. Thus, the equation of the envelope, (5.4), becomes, loosely,

$$Xe(\underline{u}) = Xf(\underline{u}) + Xs(\pm Nf(\underline{u})).$$

Along flat sections of the structuring surface, $\underline{\underline{\kappa s}}$ is singular. Therefore, the envelope would have to be parameterized by $\underline{t}$ instead of $\underline{u}$ regardless.

Note that even if both the foundation and structuring surfaces have an $n-1$-dimensional tangent space at $\underline{u}$ and $\underline{t}(\underline{u})$, the envelope may still not have an $n-1$-dimensional tangent space at $\underline{u}$. If σ is zero, then the tangents of the envelope, given by (5.13), are linearly dependent, making the normal nonexistent. This may result in a cusp or other degeneracy. Degeneracies of this nature are called "edges of regression" or "catastrophes."

5.5. Examples for Curves and Surfaces

Example 5.5.1. *Offset curves.* The first type of envelope to which we will apply these results is the offset curve. The offset curve for a given generator curve is formed by moving a circle of a given radius along the generator curve. In our notation, the foundation curve is the generator curve and the structuring curve is the circle, i.e.,

$$Xs(t) = (r\cos(t), r\sin(t))^T,$$

where r is the radius of the circle.

The normal of the structuring curve is always defined and given as

$$Ns(t) = (-\cos(t), -\sin(t))^T.$$

This makes the curvature of the structuring curve

$$\kappa s = \frac{1}{r}$$

and the curvature of the offset curve [10], [11]

$$\kappa e = \frac{\kappa f}{|1 \pm r\kappa f|}.$$

In order to find the formula for the offset curve, we solve (5.7) for t in terms of u. In this case, (5.7) is

$$\begin{pmatrix} -\cos(t) \\ -\sin(t) \end{pmatrix} = \pm Nf(u).$$

Plugging this solution into (5.4), we get the closed form for the offset curve,

$$Xe(u) = Xf(u) \pm (-rNf(u)),$$

which agrees with the standard definition of offset curves [10], [11].   □

The previous example illustrates a simple method for finding the closed form of the envelope. If it is possible to parameterize the structuring curve by the angle of its normal then $Ns(\underline{t}) = \pm Nf(\underline{u})$ may be solved easily for $\underline{t}$. This could be done if the normal of the curve is different for every parameter value. Any simple closed curve which is convex has this property where the normal is continuous. In particular, an ellipse has this property for all parameter values.

Example 5.5.2. Elliptical cutters. The curve formed by an elliptical cutting tool moving along a specified path in the plane may be calculated by using the method described for offset curves, replacing the circular structuring curve with an ellipse. Again the foundation curve is arbitrary, but the structuring curve is

$$Xs(t) = (a\cos(t), b\sin(t))^T,$$

where a and b are the lengths of the major and minor axes.

The normal of the structuring curve is always defined and given as

$$Ns(t) = \frac{1}{\sqrt{a^2\sin^2(t) + b^2\cos^2(t)}} \begin{pmatrix} -b\cos(t) \\ -a\sin(t) \end{pmatrix}.$$

This makes the curvature of the structuring curve

$$\kappa s = \frac{ab}{(a^2\sin^2(t) + b^2\cos^2(t))^{3/2}}.$$

Now, to solve (5.7) for t, we parameterize the ellipse by the angle of its normal, i.e., if θ is the angle then

$$\tan(\theta) = \frac{a}{b}\tan(t), \quad \text{or}$$

$$\sin(t) = \frac{-b\sin(\theta)}{\sqrt{a^2\cos^2(\theta) + b^2\sin^2(\theta)}}, \qquad \cos(t) = \frac{-a\cos(\theta)}{\sqrt{a^2\cos^2(\theta) + b^2\sin^2(\theta)}}.$$

This makes the equations for the ellipse

$$Xs(\theta) = \frac{1}{\sqrt{a^2\cos^2(\theta) + b^2\sin^2(\theta)}} \begin{pmatrix} -a^2\cos(\theta) \\ -b^2\sin(\theta) \end{pmatrix},$$

$$Ns(\theta) = \begin{pmatrix} \cos(\theta) \\ \sin(\theta) \end{pmatrix},$$

$$\kappa s = \frac{(a^2\cos^2(\theta) + b^2\sin^2(\theta))^{3/2}}{a^2b^2}.$$

Now to find the formula for the envelope, we use the substitution from (5.7),

$$\begin{pmatrix} \cos(\theta) \\ \sin(\theta) \end{pmatrix} = \pm Nf(u) \equiv \pm \begin{pmatrix} Nfx(u) \\ Nfy(u) \end{pmatrix}.$$

Plugging this into (5.4) we get the closed form of the envelope

$$Xe(u) = Xf(u) \pm \frac{1}{\sqrt{a^2 N f x^2 + b^2 N f y^2}} \begin{pmatrix} -a^2 N f x(u) \\ -b^2 N f y(u) \end{pmatrix},$$

whose curvature is

$$\kappa e = \frac{\kappa f}{|1 \pm (a^2 b^2 / (a^2 N f x^2 + b^2 N f y^2)^{3/2}) \kappa f|}.$$

Clearly the process becomes more complicated with more complex shapes. However, for many applications the foundation surface holds the complexity while the structuring surface is a fairly simple, smooth object. It is also often the case that the same structuring surface is used throughout the entire application while the foundation surface keeps changing. Therefore, the necessary parameterization for the structuring surface need only be computed once.

For two-parameter surface envelopes, the form of the results is basically the same as originally derived. In fact, this is true for all higher dimensions. The differential Gauss map for two-parameter surfaces is represented as a 2×2 matrix, and, therefore, the results are even more complicated than those for planar envelopes. However, there are some simple examples which illustrate key aspects of these envelopes.

Example 5.5.3. *Offset surfaces.* Offset surfaces are the counterparts to the offset curves in Example 5.5.1. The offset surface of a given generator surface is formed by moving a sphere of a given radius along the generator surface. We will use the same notation as before, but not specify exactly what parameterization we will use for the structuring surface. This will serve to avoid the inevitable singularities due to the choice of parameterization of the sphere. The only requirement for the sphere representation is that

(5.15)
$$N s(\underline{t}) = -\frac{X s(\underline{t})}{r},$$

which designates the handedness of the parameterization.

Given this form of the structuring surface, it is easy to see that

$$\underline{\underline{\kappa s}} = \frac{1}{r} \underline{\underline{I}}.$$

The similarity form in the equation for the envelope curvature matrix cancels out in this special case, and the form of the differential Gauss map for this envelope is

$$\underline{\underline{\kappa e}} = \sigma [\underline{\underline{I}} \pm r \underline{\underline{\kappa f}}]^{-1} \underline{\underline{\kappa f}}.$$

The actual formula for the envelope may be found by combining (5.7), (5.15), and (5.4) to find

$$Xe(\underline{u}) = Xf(\underline{u}) \pm (-r N f(\underline{u})).$$

Although offset surfaces have a variety of applications, they do not illustrate the full interest or complexity of more unusual surfaces. A more subtle surface is the torus, which is often used in industrial applications to cut molds for parts. The torus will serve as the main example for surface envelopes.

Example 5.5.4. *Torus drills.* A torus-shaped drill, held fixed in orientation when it is cutting, falls under the category of general offset surfaces. The foundation surface is the tool path, the envelope is the cut surface, and the structuring surface is the torus. The parameterization we will use for the structuring surface is

$$Xs(t,w) = \begin{pmatrix} (r + d\sin(w))\cos(t) \\ (r + d\sin(w))\sin(t) \\ d\cos(w) \end{pmatrix},$$

where r is distance from the center of the torus to the center of the tube, and d is the radius of the tube.

The normal of the structuring surface is always defined and given as

$$Ns(t,w) = \begin{pmatrix} -\cos(t)\sin(w) \\ -\sin(t)\sin(w) \\ -\cos(w) \end{pmatrix}.$$

This makes the curvature matrix of the structuring surface

$$\underline{\underline{\kappa s}} = \begin{pmatrix} \dfrac{\sin(w)}{r + d\sin(w)} & 0 \\ 0 & \dfrac{1}{d} \end{pmatrix}.$$

Now we need to use (5.7) to solve for $\underline{t} = (t,w)$ in terms of $\underline{u} = (u,v)$. The parameterization of the torus was chosen specifically because it matched the parameterization of a sphere with respect to its normal vector. This allows for easy inversion of $Ns(\underline{t}) = \pm Nf(\underline{u})$ once the appropriate range of w is chosen such that a single value of (t,w) is designated for any given value of (u,v). If we let w range from 0 to π, then we will be ignoring any envelopes formed by parts of the torus facing toward the center. However, it is exactly these parts of the torus that, in application, do not play a role in cutting the mold. Therefore, it is this range that we will choose for w.

Note that if the foundation normal moves across vertical, the z-axis, then there will be a discontinuity in the envelope. Again, however, this is exactly what we would expect from the drill. If the drill faced directly into the mold then it would cut a whole flat area, not just one piece of a surface.

So from (5.7) we have

$$\begin{pmatrix} \cos(t)\sin(w) \\ \sin(t)\sin(w) \\ \cos(w) \end{pmatrix} = \pm Nf(\underline{u}) \equiv \pm \begin{pmatrix} Nfx(\underline{u}) \\ Nfy(\underline{u}) \\ Nfz(\underline{u}) \end{pmatrix},$$

which implies

$$\sin(w) = \sqrt{1 - Nfz^2(\underline{u})}.$$

For these equations, we find the equation for the envelope is

$$Xe(\underline{u}) = Xf(\underline{u}) \pm \begin{pmatrix} \left[r + d\sqrt{1 - Nfz^2(\underline{u})} \right] \dfrac{Nfx(\underline{u})}{\sqrt{1 - Nfz^2(\underline{u})}} \\ \left[r + d\sqrt{1 - Nfz^2(\underline{u})} \right] \dfrac{Nfy(\underline{u})}{\sqrt{1 - Nfz^2(\underline{u})}} \\ d\,Nfz(\underline{u}) \end{pmatrix},$$

the form of the Jacobian $\underline{t}_{\underline{u}}$ is

$$\underline{t}_{\underline{u}} = \frac{1}{\sqrt{1 - Nfz^2}} \begin{pmatrix} \dfrac{Nfx\,Nfy_{\underline{u}} - Nfy\,Nfx_{\underline{u}}}{\sqrt{1 - Nfz^2}} \\ \pm Nfz(Nfx\,Nfx_{\underline{u}} + Nfy\,Nfy_{\underline{u}}) - \pm Nfz_{\underline{u}}(1 - Nfz^2) \end{pmatrix},$$

and the inverse curvature matrix is

$$\sigma\underline{\underline{\kappa e}}^{-1} = \underline{\underline{\kappa f}}^{-1} \pm \underline{t}_{\underline{u}}^{-1} \begin{pmatrix} \dfrac{r}{\sqrt{1 - Nfz^2}} + d & 0 \\ 0 & d \end{pmatrix} \underline{t}_{\underline{u}}.$$

If $\underline{\underline{\kappa f}}$ is singular then we apply (5.11) instead of (5.12). $\square$

5.6. Concluding Remarks

Offset surfaces have played an important role recently in many disciplines, particularly in CAD. However, the use of offset surfaces is limited to problems involving circles or spheres. By generalizing offset surfaces to allow for arbitrary smooth structuring surfaces, the technique may be applied to an even wider range of problems.

One such problem is that of three-axis end milling. The cutting surface for a flat end mill is a torus. Computing the tool path and verifying that path for a flat end mill is beyond the scope of conventional offset surfaces, but applies naturally to general offset surfaces. Because flat end mills are commonly used in the final stages of milling a part, general tool offsets can provide a greater level of analysis and understanding to this important process.

Further analysis and application of the methods used in this paper are warranted. Some further research is presented in the dissertation from which this paper is extracted [3], [4]. Included in that work is the application of general offset surfaces to three-axis end milling, the identification and avoidance of gouging in three-axis end milling, a discussion of general offset surfaces for foundation surfaces that are not continuous in their first derivatives, a discussion of the effects of milling considerations on part design, a further generalization of general offset surfaces to allow for the rotation and scaling of the structuring surface as it moves, the application of this further generalization to five-axis milling, and a discussion of optimal cutting paths. Further excerpts of the dissertation may be published in the future.

Methods for approximating general offset surfaces by piecewise polynomials would greatly enhance this theory and encourage its use in CAD software. General offset surfaces might also be used to compute the errors in current approximation schemes for finding and verifying tool paths for three-axis milling. In addition, general offset surfaces could be used for robot path planning and tolerance analysis.

References

[1] P. Bézier, *Numerical Control, Mathematics and Applications*, Wiley Series in Computing, John Wiley, New York, 1970.

[2] V. G. Boltyanskii, *Envelopes*, Pergamon Press, Oxford, 1964.

[3] E. L. Brechner, *Envelopes and tool-paths for three-axis end milling*, Doctoral thesis, Dept. of Mathematics, Rensselaer Polytechnic Institute, Troy, NY, 1990.

[4] _____, *General tool offset curves and surfaces*, Advances in Design Automation, 1 (1990), pp. 273–181.

[5] J. W. Bruce and P. J. Giblin, *Curves and Singularities*, Cambridge University Press, Cambridge, 1984.

[6] Y. J. Chen and B. Ravani, *Offset surface generation and contouring in Computer-Aided Design*, J. Mech. Trans. Auto. Des., 109 (1987), pp. 133–142.

[7] M. P. do Carmo, *Differential Geometry of Curves and Surfaces*, Prentice-Hall, Englewood Cliffs, NJ, 1976.

[8] L. P. Eisenhart, *A Treatise on the Differential Geometry of Curves and Surfaces*, Ginn and Co., Boston, 1909.

[9] R. T. Farouki, *The approximation of non-degenerate offset surfaces*, Comput. Aided Geom. Des., 3 (1986), pp. 15–43.

[10] R. T. Farouki and C. A. Neff, *Algebraic properties of plane offset curves*, Comput. Aided Geom. Des., 7 (1990), pp. 101–127.

[11] _____, *Analytic properties of plane offset curves*, Comput. Aided Geom. Des., 7 (1990), pp. 83–99.

[12] I. D. Faux and M. J. Pratt, *Computational geometry for design and manufacture*, in Mathematics and its Applications, John Wiley & Sons, New York, 1979.

[13] H. W. Guggenheimer, *Differential Geometry*, Dover Publications, Inc., New York, 1977.

[14] J. Hoschek, *Offset curves in the plane*, Comput. Aided Des., 17 (1985), pp. 77–82.

[15] _____, *Spline approximation of offset curves*, Comput. Aided Geom. Des., 20 (1988), pp. 33–40.

[16] J. Hoschek and N. Wissel, *Optimal approximate conversion of spline curves and spline approximation of offset curves*, Comput. Aided Des., 20 (1988), pp. 475–483.

[17] R. Klass, *An offset spline approximation for plane cubic splines*, Comput. Aided Des., 15 (1983), pp. 297–299.

[18] E. P. Lane, *Metric Differential Geometry of Curves and Surfaces*, The University of Chicago Press, Chicago, 1940.

[19] S. G. Papaioannou and D. Kiritsis, *An application of Bertrand curves and surfaces of CAD/CAM*, Comput. Aided Des., 17 (1985), pp. 348–352.

[20] B. Pham, *Offset approximation of uniform B-splines*, Comput. Aided Des., 20 (1988), pp. 471–474.

[21] M. H. Protter and C. B. Morrey, *Implicit function theorems and differential maps*, in A First Course in Real Analysis, Springer-Verlag, New York, 1988, pp. 332–364.

[22] B. Ravani and T. S. Ku, *Bertrand offsets of ruled surfaces*, Comput. Aided Des., 1990, to appear.

[23] D. J. Struik, *Lectures on Classical Differential Geometry*. Dover Publications, Inc., New York, 1961.

[24] W. Tiller and E. G. Hanson, *Offsets of two-dimensional profiles*, IEEE Comput. Graphics Appl., 4 (1984), pp. 36–46.

Surface-Surface Intersection

Adaptive Contouring for Triangular Bézier Patches

Robert E. Barnhill, Brett K. Bloomquist, and Andrew J. Worsey

6.1. Introduction

Contouring is an often used and important approach to the problem of visualizing and interrogating surfaces. Even with the present sophistication of interactive color graphics for displaying bivariate surfaces, it is needed to examine the reflection lines of a bivariate function [5]. Reflection lines offer a useful technique for interrogating a surface and are used extensively in the car industry for judging the aesthetic quality of a surface [8].

Most contour methods fall into one of two basic categories: "grid-type" or "gridless" methods. The approaches taken in these two cases are markedly different and lead to algorithms with, correspondingly, different strengths and weaknesses. References to, and a discussion of, such schemes may be found in [10] and [2].

An adaptive contouring method which combines the strengths of these two approaches was developed by Petersen [11]. It contours a piecewise polynomial surface defined as a collection of triangular Bézier patches, such as that obtained, for example, by using either the Powell–Sabin or Clough–Tocher interpolation schemes for scattered data defined over a triangulation in $I\!R^2$ [13], [7]. The algorithm proceeds by implementing a scheme of degree reduction combined with subdivision to generate a local piecewise linear approximation to within some prescribed tolerance of the original surface. It is then a straightforward matter to obtain piecewise linear approximations to the required contour level. The adaptive nature of the algorithm ensures that, to within tolerance, no contours are missed and since the degree reduction and subdivision are only applied locally in the region of the contour, the method is reasonably fast and efficient.

Given these strengths, we take Petersen's algorithm as the basis for developing a better scheme for contouring triangular Bézier patches. The algorithm we present in this paper is more robust and efficient than Petersen's and the improvement comes from changing only three basic ideas in his method.

First of all, we propose a superior method for degree reduction than the simple averaging technique used in [11]. The method is based on least squares approximation and greatly reduces the number of domain subdivisions that are needed in order to locally approximate the original surface to within tolerance.

Secondly, we terminate the degree reduction process once we have generated a piecewise quadratic approximation to the original surface. We then contour this *exactly* using the algorithm of Worsey and Farin [15]. Therefore, contours are approximated by piecewise conic sections which, depending on the original surface and the affect of the degree reduction, might possibly be C^1. This contrasts sharply with Petersen's approach, which is to degree reduce down to a piecewise linear approximation before contouring. That is generally far more expensive computationally, because more subdivisions are usually required in order to obtain an approximation to within tolerance. It also means that contours are approximated by piecewise linear curves which, except in completely pathological cases, are C^0 at best.

Finally, we further modify Petersen's algorithm by implementing a different strategy for ordering domain triangles that have been subdivided and need to be processed further. This has the effect of reducing the total number of triangular patches that ultimately have to be contoured, leading to further computational savings and improved numerical stability.

Since it is the basis for our method, we give a complete outline of Petersen's method in the next section, before discussing the above improvements we have implemented in more detail. That is done in §6.3, where we present a full description of our algorithm. We conclude the paper with some graphical results in §6.4, showing the contours generated by this new algorithm for certain test cases.

6.2. Petersen's Contouring Algorithm

Petersen's algorithm [11] is designed to contour a piecewise polynomial function defined as a collection of triangular Bézier patches of degree n. The contouring algorithm revolves around five basic steps.

1. Find a triangular patch that possibly contains the contour level, culling those that cannot by using the convex hull test for a Bézier patch.

2. If $n = 1$, go to step 5. If not, approximate the nth degree patch with an $(n-1)$st degree patch.

3. Check to see if the $(n-1)$st degree patch approximates the nth degree patch to within a user-specified tolerance. If so, replace n by $(n-1)$ and go to step 2, else:

4. Split the domain triangle its neighbor (so as to maintain a valid triangulation), as well as the polynomial surface patches over these triangles, and continue at step 2 for each new patch that possibly contains the contour level.

5. Contour the linear patch and draw the contour segment. If there are more patches to contour, go back to step 1.

Typically, the data structure for a triangle in the triangulation contains pointers to the data structures for neighboring triangles. This conveys information about interconnection topology of the triangulation, so that after finding an initial patch to contour (step 1), we can make use of the neighbor information to find the next patch to contour.

Complete details of each of the five steps of the algorithm are covered in [11]. However, it is relevant to consider them again here in some detail, in order to see how they impact the contouring algorithm and how they might be improved. The modifications that are implemented in the new scheme in §6.3, amount to changes in all of these steps. Steps 2 and 3 stand alone and are considered first. The remaining three steps are very much related, dictating, as they do, the order in which the domain triangles are processed in the contouring algorithm.

6.2.1. Step 2: Degree Reduction.

Reducing the degree of a patch from n to $(n-1)$ is an approximation scheme. Petersen [11] does this by first of all using the (exact) degree elevation formula for expressing a triangular Bézier patch of degree $(n-1)$ as one of degree n [3], [6], [7]. Rewriting the degree elevation formulae, one obtains the following overdetermined linear system for the coefficients b_λ, $|\lambda| = n - 1$, of the degree $(n-1)$ patch in terms of those, $\bar{b}_\mu$, $|\mu| = n$, of the degree n patch,

$$(6.1) \qquad \frac{1}{n} \sum_{i=1}^{3} \mu_i b_{\mu - \epsilon_i} = \bar{b}_\mu, \qquad |\mu| = n,$$

in which the standard multi-indexing notation is used [6]. In order to maintain interpolation at the vertices, the constraints

$$(6.2) \qquad b_{n-1,0,0} = \bar{b}_{n,0,0}, \quad b_{0,n-1,0} = \bar{b}_{0,n,0}, \quad \text{and} \quad b_{0,0,n-1} = \bar{b}_{0,0,n},$$

are necessarily imposed.

Petersen "solves" this system by using a simple averaging technique. Starting at any one of the three vertices of the triangle, he uses a forward differencing scheme to obtain some of the Bézier ordinates b_λ. (Precisely which ordinates these are is discussed in [11].) The method is then applied cyclically to each of the other two vertices in turn, in order to define the remaining ordinates and thereby generate the degree $(n-1)$ approximation to the (given) degree n surface patch.

This degree reduction scheme is, of course, simple, but it is far from optimal in any approximation theoretic sense. Indeed, the degree $(n-1)$ patch is often a very poor approximation of the original patch and this can prove to be computationally expensive, since the patch must then be subdivided (in step 3).

In our algorithm we use a degree reduction scheme based on a least squares solution to the overdetermined linear system (6.2), as discussed in §6.3. This approach has a firm mathematical underpinning and produces a better approximation to the original degree n patch.

6.2.2. Step 3: Testing the Approximation. A decision must be made as to whether or not the lower degree approximation is acceptable. This is done by first of all raising the degree of the approximation from $(n-1)$ to n [3], [6] to generate the Bézier ordinates $b_\lambda, |\lambda| = n$. These are then compared to those, $\bar{b}_\lambda$, of the original patch. Specifically,

$$\sum_{|\lambda|=n} (b_\lambda - \bar{b}_\lambda)^2$$

is used as a measure of how well the lower degree patch approximates the original, by comparing it to a prescribed tolerance ϵ. If

$$\sum_{|\lambda|=n} (b_\lambda - \bar{b}_\lambda)^2 < \epsilon,$$

the degree reduction is satisfactory and we move on to consider another triangle. If not, the current triangle is subdivided.

6.2.3. Steps 1, 4, and 5: Processing the Triangles. All of these steps affect, in some way, the precise order in which triangles are processed for contouring. Step 1 is a simple culling process that improves efficiency of the algorithm. From the convex hull property of Bézier patches [3], a triangle need not be considered for contouring if the contour level is outside the convex hull of the Bézier ordinates defining the surface. This test is used at the beginning of the algorithm but can be implemented at a later stage as well, after triangles in the domain have been split.

To split a triangle in the domain, the midpoint of the longest edge is joined to the opposite vertex. This strategy ensures that in *that* triangle the smallest angle is never reduced, leading to certain desirable results [14]. In particular, it helps in avoiding "skinny" triangles although such avoidance cannot be guaranteed since, in order to maintain a valid triangulation and C^0 continuity between patches, the neighboring triangle on the edge is also subdivided by joining the midpoint to the opposite vertex.

As implemented, the splitting algorithm is applied locally to an initial triangle (and therefore, with iteration, subtriangles of the same) until a linear patch approximating the original is obtained. The contour segment over this patch is then drawn, ostensibly to facilitate drawing contours on a pen plotter. Consequently, the splitting process is not applied in tandem with degree reduction, and this separation of the two procedures induces two major problems.

First, the algorithm may repeatedly subdivide the surface in a local region, leading to a very fine mesh in some parts of the domain. This can (and does) lead to difficulties when rendering the piecewise linear approximation to the contour. A more reasonable approach is to process the domain triangles using more of a "breadth-first" approach, based on areas.

Second, the way the triangular patches are processed does not in fact *guarantee* even C^0 contour levels. This is because the contour segment is being drawn *during* the contouring process. To illustrate the point, suppose that we have successfully degree reduced a quadratic patch to a linear patch, have, therefore, drawn the contour segment, and are now contouring a neighboring patch. If the degree reduction step is not successful, this patch is split. In doing this, it is entirely possible that the patch will be subdivided along the edge common to the neighboring triangle that has just been contoured. This produces a discontinuity (tear) between the patches, and a corresponding discontinuity in the contour level.

Both of these problems arise because of the way the subdivision algorithm in step 4 is systematically applied moving from one region of the surface to another. It has nothing to do with the split on a local level, but rather is a global effect. Furthermore, the algorithm degree reduces a patch down to a linear approximation before contouring. This is clearly not necessary since a quadratic triangular Bézier patch can be contoured *easily* and *exactly* [15]. Doing so leads to significant computational savings since the degree reduction from a quadratic to a linear patch in steps 2 and 3 is invariably unsuccessful, thereby enforcing several levels of subdivision for a patch in step 4.

In the method presented in the next section, we change these aspects of Petersen's approach [11], to develop a more stable, efficient, and robust contouring algorithm. Although the method is essentially based on a collection of enhancements to Petersen's work, it does stand alone as a new approach to the contouring problem.

6.3. A New Contouring Algorithm

Our new method has the following basic steps.

1. Cull all triangles, by using the convex hull test for Bézier patches, that cannot contain the contour level and then find a patch of degree n that may contain the contour. If there are no patches of degree n, replace n by $(n-1)$. If $n = 2$, go to step 5.

2. Approximate the nth degree patch with an $(n-1)$st degree patch, using a least squares approximation.

3. Check to see if the $(n-1)$st degree patch approximates the nth degree patch to within a user-specified tolerance. If it does, move the patch to a list of degree-reduced patches (this does not affect the interconnection topology of the triangulation in any way) and go to step 1.

4. Split the domain triangle in a way that will maintain a valid triangulation. Go to step 2.

5. Contour the *quadratic* approximation over the triangle using the method of [15].

Fundamentally, this is, as we have said, the same approach used in [11]. There are, however, significant changes, on both a theoretical and implementation level. We now consider the details of these steps in turn, indicating why they lead to an improved method for contouring triangular Bézier patches.

6.3.1. Steps 1 and 4. The essence of the approach in step 1 is to degree reduce (when appropriate) the surface *globally*, rather than working on one local area at a time. It is important to note, however, that patches not containing the contour are being removed from the contouring process. So, computational effort is only being spent where it is needed.

Petersen [11] starts to contour on the domain triangle with the largest area. He found that this produced a numerically stable starting point. However, the measure of area is only used to start with. In our approach, we always work on the domain triangles having the largest area when we find a patch of degree n that may contain the contour. This means our algorithm follows a *breadth first* approach rather than the *depth first* method used by [11], leading to a far more stable algorithm.

In working on triangles of largest area, we see that step 1 is affected by step 4, in which the splitting takes place. The split we use for a domain triangle is precisely the same as that used by [11], namely, to split a triangle in the domain, the midpoint of the longest edge is joined to the opposite vertex and, in order to maintain a triangulation, the vertex of the opposite, neighboring triangle as well. In order to correlate the two and keep track of the processing order, we implement the following idea.

First, find the triangle of largest area and assign it an index of 0. Then, assign all remaining triangles an index according to the formula

$$index = \left\lfloor \log_2 \left(\frac{\text{area } (largest\ triangle)}{\text{area } (current\ triangle)} \right) \right\rfloor.$$

When a triangle is split at the midpoint of its longest edge (or any edge for that matter), its area is halved. Therefore, with the above definition for the index of a triangle, when we split a domain triangle, we have only to increment the index by one. In processing the triangles for contouring, we first of all work on all triangles with index 0, then all triangles with index 1, and so on, until all of them can be degree reduced to within tolerance.

This approach has yielded excellent results and can be implemented efficiently using dynamic data structures. It has also been used for trivariate contouring where the improvement is even more pronounced [1].

6.3.2. Step 2. If we again consider degree reduction as the inverse process to degree elevation, we have to solve the overdetermined system (6.1). Since we want to approximate the patch of degree n by one of degree $(n - 1)$ in some optimal way, it makes far more sense to "solve" this system by using least squares approximation, rather than the ad hoc averaging technique implemented by [11]. This can be done very easily by solving a linear system [4], and gives far better results. Since our method generates better approximations, it also means that less domain subdivision is usually necessary.

However, one cannot simply solve the least squares problem without regard to the overall problem. The degree reduction process cannot be implemented in isolation, over one triangle at a time. Care must be taken to maintain continuity of the surface; otherwise the contouring algorithm will fail. Therefore, we must ensure that the degree reduction on an edge depends only on data on that edge. We do this by solving the univariate least squares degree reduction problem for each of the boundary curves first of all, and then modify it so as to maintain interpolation at the vertices, that is, we impose the constraints (6.2). Having degree reduced the boundary curves, we then solve for the *remaining* interior Bézier points in (6.1) by obtaining the least squares solution to the (reduced) overdetermined linear system.

6.3.3. Step 3. It is important to note that the test Petersen uses for degree reduction, namely

$$\sum_{|\lambda|=n} (b_\lambda - \bar{b}_\lambda)^2 < \epsilon,$$

depends only on the Bézier ordinates, and not on the "size" of the domain triangle. In our method we have modified this, improving performance. Specifically, if A is the area of the domain triangle under consideration, we consider the degree reduction acceptable if

$$\sum_{|\lambda|=n} A \cdot (b_\lambda - \bar{b}_\lambda)^2 < \epsilon.$$

This is not the most significant change of those we implement but it does, nonetheless, improve the algorithm.

6.3.4. Step 5. When the algorithm enters this step, the patches that might contain the contour level have all been degree reduced to a quadratic approximation. At this stage we simply implement the algorithm of [15] for contouring a quadratic triangular Bézier patch *exactly*. That algorithm is simple, efficient, and robust. Using it, we ultimately approximate the original contour level as a collection of piecewise rational quadratic Bézier curves.

This approach is *significantly* different from that taken by Petersen in [11], who, as we have seen, degree reduces patches down to a linear approximation. This step alone makes our algorithm far more efficient, since considerably less domain subdivision is required.

6.4.   Examples and Conclusions

We have presented an algorithm for contouring a function defined as a collection of triangular Bézier patches. The algorithm is based on an earlier one due to [11] but incorporates several important changes that greatly improve the method. It is fast, efficient, robust, and ultimately approximates contours by a piecewise rational quadratic Bézier curve (conic sections), to within a user-specified tolerance. Depending on the original surface and the effects of the degree reduction, the piecewise defined curve may be globally C^1 continuous.

We have tested the contouring algorithm on a number of examples but, for the sake of brevity, only include the results from four. Other results are given in [2]. In each example we consider a different function, but each is defined over the same triangulation, T, obtained by forming the Delaunay triangulation of the 25 uniformly gridded data sites

$$(6.3) \qquad (i/4, j/4) \,, \quad i \in 0, 1, \cdots, 4, \quad j \in 0, 1, \cdots, 4.$$

Consequently, the domain of interest is $[0, 1] \times [0, 1]$.

The functions in the four examples are all generated as the C^1 piecewise cubic Clough–Tocher interpolant with linearized cross-boundary derivatives (see [5]), to certain positional and gradient data at the vertices (6.3) of the triangulation T. It is these data that change in the examples.

Example 1. We take positional data from the function

$$F(x, y) = (x - 0.5)^2 + (y - 0.5)^2,$$

and use the triangular Shepard's method [9] to estimate gradients. Figure 6.1 illustrates the results of contouring the Clough–Tocher interpolant to these data. Contour levels for the ten function values ranging from 0.0 to 0.45 in steps of 0.05 are shown (although the zero contour level is simply the origin), using a tolerance of 10^{-6}. This is one of the examples considered by [11], and the results clearly demonstrate the considerable improvement gained by using our algorithm. In this case, this is primarily because we contour at the quadratic level of approximation as opposed to linear.

Example 2. We again take positional data from the function

$$F(x, y) = (x - 0.5)^2 + (y - 0.5)^2,$$

but now use *exact* gradient data to generate the Clough–Tocher interpolant. Since the interpolant has quadratic precision, it is in fact equivalent to F, and because our algorithm contours quadratics exactly, the contours generated are circles; see Fig. 6.2. Consequently, no degree reduction is involved in contouring this example, which is included primarily for comparison with the results from Example 1. As before, we calculate contours for the ten function values ranging from 0.0 to 0.45 in steps of 0.05.

Example 3. We take positional data from the function

$$F(x, y) = 2[(x - 0.5)^3 + (y - 0.5)^3],$$

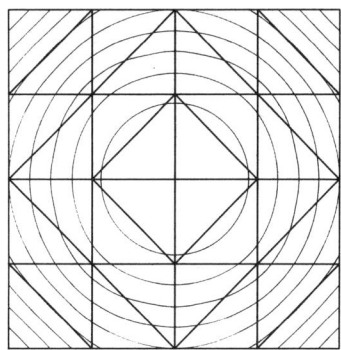

FIG. 6.1. *Ten contour levels are shown for the Clough–Tocher interpolant generated using data from Example 1. The initial triangulation of the 25 data sites is also shown.*

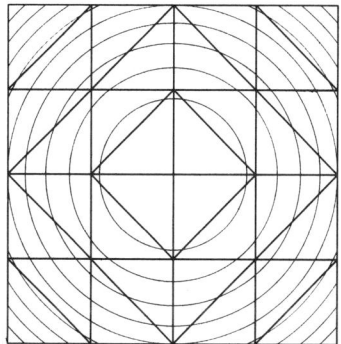

FIG. 6.2. *Ten contour levels are shown for the Clough–Tocher interpolant generated using the data from Example 2. Since the interpolant is in fact a global quadratic, the circular contours are computed exactly.*

and use the triangular Shepard's method [9] to estimate gradients. Figure 6.3 illustrates the results of contouring the Clough–Tocher interpolant to these data. Contour levels for the 19 function values ranging from 0.025 to 0.475 in steps of 0.025 are shown and are computed using a tolerance of 10^{-8}. The "wavy" nature of the contours is due to the gradient estimation scheme used. It has nothing at all to do with the contouring algorithm which, because of the very small tolerance, computes the contours exactly to within machine precision. This is confirmed by the results of Example 4.

Example 4. We take positional data from the function

$$F(x,y) = 2[(x - 0.5)^3 + (y - 0.5)^3],$$

but use *exact* gradient data to generate the Clough–Tocher interpolant. As with Example 3, we compute contours of the interpolant for the 19 function values ranging from 0.025 to 0.475 in steps of 0.025, using a tolerance of 10^{-8}. They are shown in Fig. 6.4.

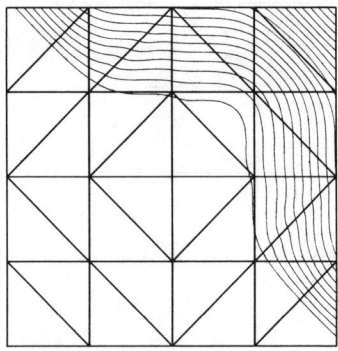

FIG. 6.3. *Nineteen contour levels for the Clough–Tocher interpolant generated using data from Example 3. The initial triangulation of the 25 data sites is also shown.*

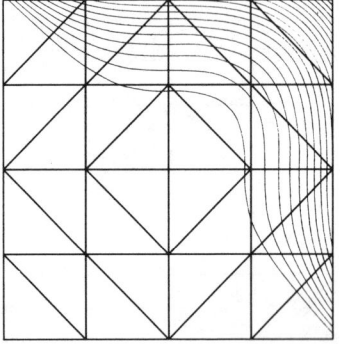

FIG. 6.4. *Nineteen contour levels for the Clough–Tocher interpolant generated using data from Example 4. The initial triangulation of the 25 data sites is also shown.*

These and other results clearly demonstrate the superiority of our contouring algorithm over that of Petersen [11], upon which it is based. The improvement comes from making certain key changes in the details of Petersen's method. That said, his work gave us an excellent base to build upon and his original ideas for contouring are not only sound, but lend themselves to developing efficient and robust algorithms for contouring piecewise polynomial functions.

We have extended several of the ideas presented here to the problem of contouring trivariate Bézier patches [1]. That work is based on an earlier algorithm developed in [12]. The contouring problem for trivariate functions is extremely important, since it is not possible to analyze such surfaces directly by graphical means, and the advantages that our new approach offers in the bivariate case are more pronounced in three dimensions.

Acknowledgments

This research was supported in part by Department of Energy grant DE-FG02-87ER25041 and National Science Foundation grant DMC-8807747, both awarded to Arizona State University, Tempe, AZ, as well as National Science Foundation grant DMS-8803257, awarded to the University of North Carolina at Wilmington, NC. The authors thank Nathan Watson for his help in producing the figures.

References

[1] R. E. Barnhill, B. Bloomquist, and A. J. Worsey, *Adaptive contouring for trivariate Bézier surfaces*, 1991. In preparation.

[2] B. Bloomquist, *Contouring trivariate surfaces*, Master's thesis, Department of Computer Science, Arizona State University, Tempe, AZ, 1990.

[3] W. Boehm, G. Farin, and J. Kahmann, *A survey of curve and surface methods in CAGD*, Comput. Aided Geom. Des., 1 (1984), pp. 1–60.

[4] E. W. Cheney, *Introduction to Approximation Theory*, McGraw-Hill, New York, 1966.

[5] G. Farin, *A modified Clough–Tocher interpolant*, in Surfaces in Computer Aided Geometric Design '84, R. E. Barnill and W. Boehm, eds., North-Holland, Amsterdam, 1985, pp. 19–27.

[6] ———, *Triangular Bernstein–Bézier patches*, Comput. Aided Geom. Des., 3 (1986), pp. 83–127.

[7] ———, *Curves and Surfaces for Computer Aided Geometric Design*, second edition, Academic Press, New York, 1990.

[8] R. Klass, *Correction of local surface irregularities using reflection lines*, Comput. Aided Des., 12 (1980), pp. 73–77.

[9] F. F. Little, *Convex combination surfaces*, in Surfaces in Computer Aided Geometric Design, R. E. Barnhill and W. Boehm, eds., North-Holland, Amsterdam, 1983, pp. 99–107.

[10] C. Petersen, *Contours of three and four dimensional surfaces*, Master's thesis, Department of Mathematics, University of Utah, Salt Lake City, UT, 1983.

[11] ———, *Adaptive contouring of three dimensional surfaces*, Comput. Aided Geom. Des., 1 (1984), pp. 61–74.

[12] C. Petersen, B. Piper, and A. J. Worsey, *Adaptive contouring of a trivariate interpolant*, in Geometric Modeling: Algorithms and New Trends, G. Farin, ed., Society for Industrial and Applied Mathematics, Philadelphia, PA, 1987, pp. 385–395.

[13] M. J. D. Powell and M. A. Sabin, *Piecewise quadratic approximation on triangles*, ACM Trans. on Math. Software, 3 (1977), pp. 316–325.

[14] I. G. Rosenberg and F. Stenger, *A lower bound on the angles of triangles constructed by bisecting the longest side*, Math. Comp., 29 (1975), pp. 390–395.

[15] A. J. Worsey and G. Farin, *Contouring a bivariate quadratic polynomial over a triangle*, Comput. Aided Geom. Des., 7 (1990), pp. 337–351.

Constructive Geometric Approach
to Surface-Surface Intersection

Les A. Piegl

7.1. Introduction

Surface-surface intersection, commonly referred to as SSI, has been the subject of considerable interest of computer aided design (CAD) since the early 1960s. Several algorithms have been developed for such purposes as surface trimming, blend and fillet surface generation, offsetting, cutter path generation, and Boolean operations for solids' modeling. It is amazing (and annoying) to see that the SSI problem, as old and as well researched as it is, does not have a sufficiently robust and reliable solution acceptable in CAD applications. Most of the problems are numerical and topological in nature, e.g., it is difficult to establish and understand the topology of the intersection curve(s) and to make sure that no part of the intersection is missed and traced out with acceptable accuracy. This paper introduces a method which has folklore roots in constructive/descriptive geometry [1]–[6]. Before the advent of computers, intersections were constructed by hand relying on the engineer's imagination and training in three-dimensional (3D) construction techniques (Fig. 7.1 illustrates the idea). It turns out, however, that these methods are easily computerized and, what is more important, they work extremely well *on the computer*. Based on the limitations of constructive geometry, the following surface types can be considered:

- natural quadrics (plane, cone, cylinder, and the sphere),
- line/circular arc extrusion,
- revolved line/circular arc, and
- general extruded surface.

The reasons why these surface types are considered special can be summarized as follows:

- They appear very frequently in engineering design; in fact, the majority of mechanical objects can be designed by natural quadrics [7]–[10].
- It is relatively easy to find the intersection curves and to understand their topology.
- Special cases such as touch points or overlapping can be handled very reliably.

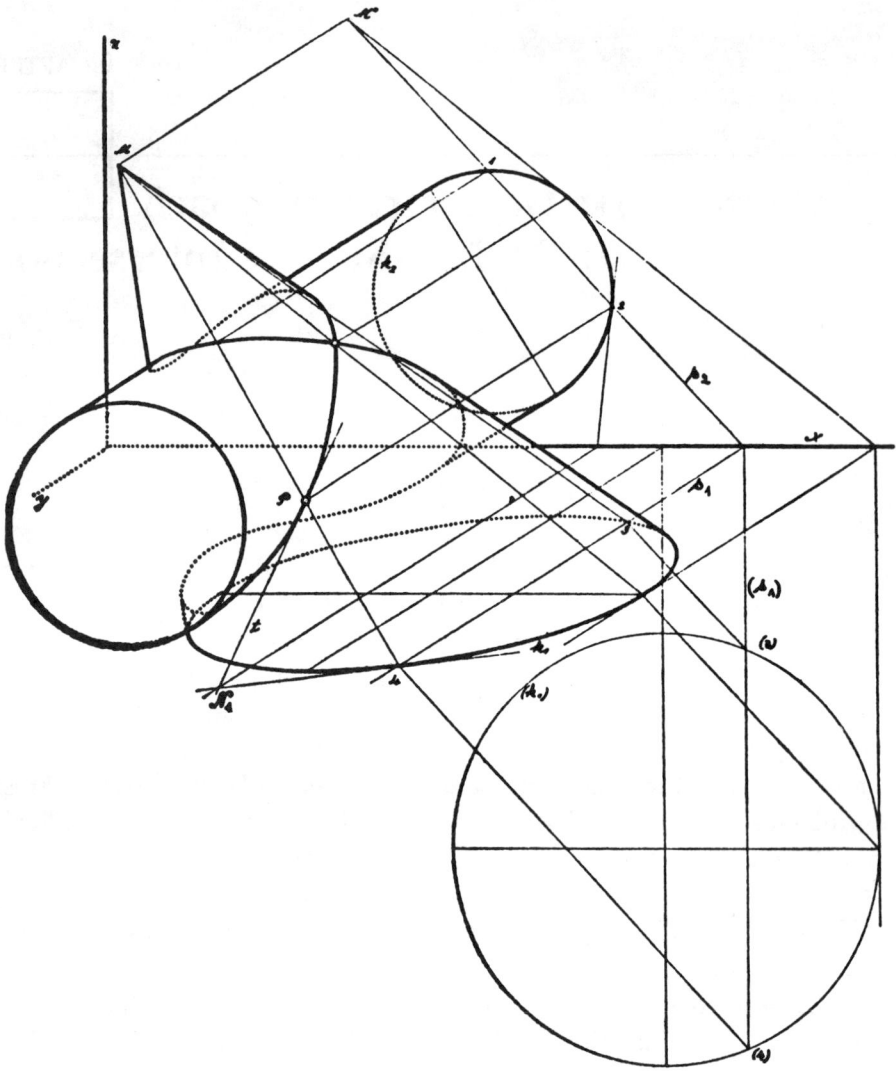

FIG. 7.1. *Classical constructive geometric method used to construct surface-surface intersection by hand.*

- Robust geometric methods can be applied to obtain intersection points. These methods use virtually no numerical and algebraic techniques and hence are not subject to numerical failure [11].

In the sections that follow, we outline techniques that compute the intersection curves among the above-mentioned surfaces represented in trimmed nonuniform rational B-spline (NURBS) form [12], i.e., in tensor product form. Although the techniques are based on traditional constructive geometric methods, each surface pair has to be considered separately as the topology and the type of intersection depend on the surfaces being intersected.

7.2. Computing Surface Intersections

Given two trimmed NURBS surfaces that represent the objects discussed above, the algorithm computes the model space and/or the parameter space representations of the intersection curves that satisfy the following requirements:

- In all three spaces the intersection curve must have the same number of segments.
- The orientation of the three representations of the same curve must be the same.
- All curves must be found including the just touch cases.

These requirements are imposed by the Boolean operator that works in the parameter spaces and assumes proper orientation and segmentation of the intersection curves [for simple applications, e.g., blending (not necessarily parameter space driven), where only the model space curves are needed, the above requirements can be relaxed substantially]. The intersection process consists of the following major steps [13]:

1. Determine the type of intersection.
2. Understand the topology of the intersection curve.
3. Detect and compute conic sections.
4. Compute general intersection points.
5. Include break points.
6. Handle special cases.
7. Clip the intersection curves.
8. Link the segments, and
9. Represent intersection curves in all three spaces.

7.2.1. Type of Intersection.
There is no general recipe as to how to figure out the type of intersection. The reason for that is the fact that each pair of surfaces results in different types of intersection. However, if one considers each pair of surfaces separately, it is not too difficult to establish how many and what kind of intersection curves are obtained based on experience and/or imagination. For illustrative purposes, we consider here the cylinder/sphere case. The types of intersection that can occur are illustrated in Fig. 7.2. They are: (1) nonintersection, (2) touch point, (3) one object bites into the other, (4) the cylinder touches the sphere from inside, and (5) the cylinder penetrates the sphere. Each case can be further classified according to the relative positions of the two objects, e.g., the penetrate case can result in one or two circles if the two objects share the same axis of symmetry.

7.2.2. Understanding the Topology.
Within each type of intersection the topology of the intersection curve has to be understood in order to be able to compute a sequence of points that sweep out the curve in a proper order. Again, there is no ultimate method of obtaining a topological classification.

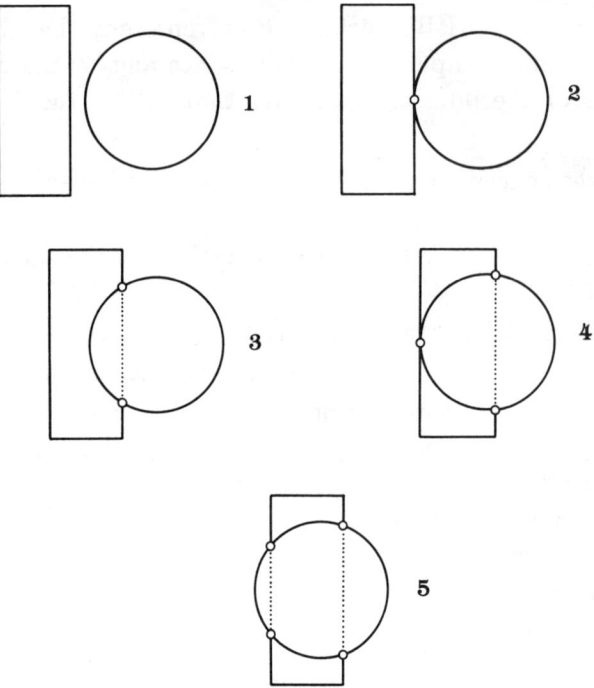

FIG. 7.2. *Classification of intersection types in case of cylinder-sphere intersection.*

It is the implementor who figures out the curve's shape *before* intersection points are computed as it is absolutely necessary to know how many curves we have, and within each curve whether the curve is open or closed, and if it is closed, then whether it has singular points or not, etc. A simple classification is illustrated in Fig. 7.3, where the torus is intersected with a plane parallel to its axis. Based on topological reasoning and simple imagination, the following intersection classification can be obtained: (1) two closed curves, (2) one closed curve with a double point, (3) one closed curve with four inflection points, and (4) one closed curve without an inflection point.

7.2.3. Detecting Conic Intersection.   In engineering design, conic sections play an important role. Therefore it is important to detect conic intersections and compute them *precisely.* Conic intersections arise during plane-quadric, quadric-quadric, and in special cases such as the plane-torus.

The plane quadric case is covered by most of the engineering geometry books and therefore it is not discussed in this paper (Fig. 7.4(a) illustrates the plane-cone case). Quadric-quadric intersection can result in conic sections in two cases: (1) the two objects are symmetric with respect to the same axis, such as the one illustrated in Fig. 7.4(b), and (2) the surfaces have two intersection points where the tangent planes are the same and have at least two other intersection points such that these four points are not coplanar (Fig. 7.4(c))

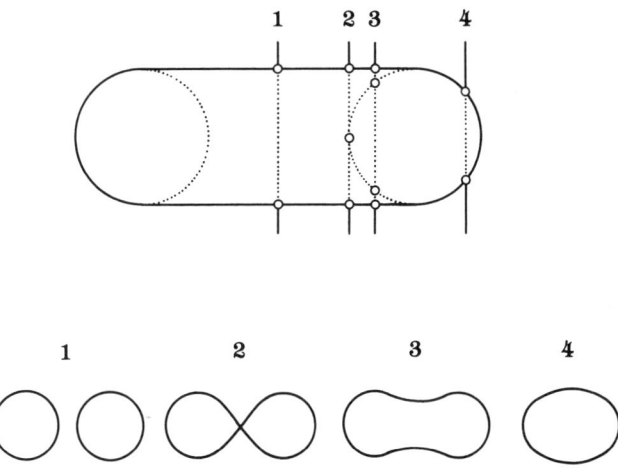

FIG. 7.3. *Topological classification of plane-torus intersection when plane is parallel to the axis of the torus.*

[14]–[16]. To detect such intersection in case of the cone and the cylinder, the following simple algorithm can be used (refer to Fig. 7.5):

```
detect_conic(cone,cylinder,conic?)
{
  conic?=0;
  {i1,i2}=intersect(planes{PY,PO},line[A,V]);
  {ty1,ty2}=tangents(i1,bottom circle of cylinder);
  {to1,to2}=tangents(i2,bottom circle of cone);
  m=intersect(PO,PY);
  {BY,TY}=intersect(m,{ty1,ty2});
  {BO,TO}=intersect(m,{to1,to2});
  if(abs(BO-BY)<TOL && abs(TO-TY)<TOL) conic?=1;
}
```

The shorthand notation {i1,i2}=intersect(planes{PY,PO}, line[A,V]) is interpreted as follows: planes PY and PO are intersected with the line AV yielding two intersection points i1 and i2. Other notational conventions found in subsequent algorithms can be interpreted similarly.

The computational costs are line-plane and line-line intersections and the computations of tangents to circles. These can be done extremely fast and accurately as all the computations are performed in a local coordinate system. The only tolerance used in the algorithm is TOL which is the point coincidence tolerance. Therefore it is not subject to numerical failures that are usually the result of inconsistent tolerancing. The above algorithm can be modified easily to handle cone-cone and cylinder-cylinder intersections, and the ideas can be carried over to other cases such as cylinder-ellipsoid.

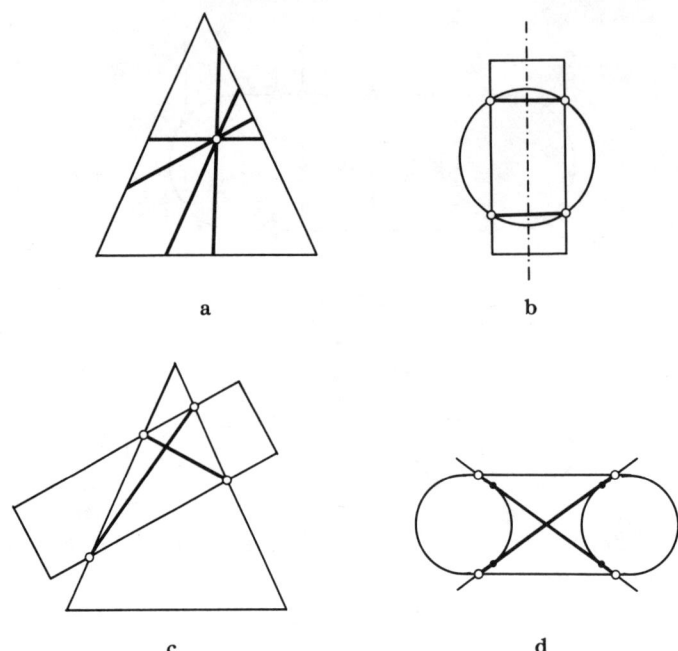

FIG. 7.4. *Types of conic intersection:* (a) *plane-quadric,* (b) *sharing axis of symmetry,* (c) *quadric-quadric having two double points,* and (d) *plane tangential to torus at two hyperbolic points.*

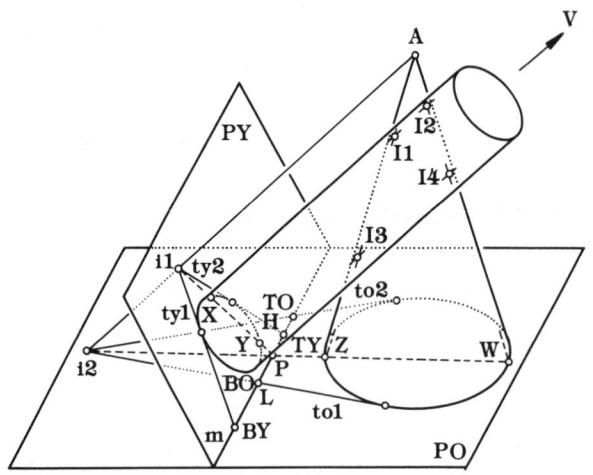

FIG. 7.5. *Slicing technique applied to compute series of intersection points.*

Conic intersections can arise in case of higher degree surfaces such as the torus which is of practical interest. Cutting the torus with a plane which is tangential to the torus at two points (see Fig. 7.4(d)), yields an intersection curve that consists of two circles (these circles are called *Yvone-Villarceau* circles [17]). Since the torus/toroidal patch appears frequently in such applications as rounding and the generation of surfaces of revolution, the above plane intersection case is of practical importance.

7.2.4. Computing Intersection Points.

The general technique, imported from traditional descriptive geometry [1]–[6], is slicing: slice the two objects with planes such that the resulting intersection curves are simple, i.e., lines and circles, and thus the surface intersection points are obtained by intersecting these planar curves. The crucial point is how to organize the slicing so that it will be implementable on the computer. As each case is different, there is no generally acceptable method. However, there are a few rules of thumb that can be considered: (1) for ruled surfaces, e.g., cones and cylinders, slice the objects with planes that cut both surfaces in lines; (2) if one of the objects is a sphere, then use planes that cut the other surface in simple curves as each cut of the sphere is a circle; (3) if surfaces of revolution are involved, it is a good idea to cut with planes that are perpendicular to the axis of revolution. We consider an example for case (1) and show how it can be computerized using the cone-cylinder intersection discussed above. Referring to Fig. 7.5, the algorithm goes like this:

```
compute_intersection(cone,cylinder,k,I[])
{
  get BO, TO, BY and TY as in the previous algorithm;
  if([BO,TO]∩[BY,TY]=∅) { No intersection;
exit(1); }
  [L,H]=[BO,TO]∩[BY,TY];
  delta=1.0/(k-1); n=4*(k-2)+5; half=(n+1)/2;
  for(i=1; i<=k; i++)
  {
    u=delta*(i-1);
    P=(1-u)*L+u*H;
    {X,Y}=intersect(line[i1,P],cylinder's bottom circle);
    {Z,W}=intersect(line[i2,P], cone's bottom circle);
    j1=i; j2=n-i+1; j3=half-i+1; j4=half+i-1;
    I[j1]=intersect(line[A,Z],line[V,X]);
    I[j2]=intersect(line[A,W],line[V,X]);
    I[j3]=intersect(line[A,Z],line[V,Y]);
    I[j4]=intersect(line[A,W],line[V,Y]);
  }
}
```

This algorithm assumes that an a priori analysis has been made and determined the minimum number of slices (k in the above algorithm) that result in sufficient number of intersection points. In general, the slicing results in four points of intersection, and therefore if k slices are applied, then $4(k-2)+5$ intersection points are computed. Introducing the variable half $= (n+1)/2$, the four points can be indexed as i, $n - i + 1$, half $- i + 1$ and half $+ i - 1$. If the slicing plane moves from the position L to the position H, then the above indexing results in a continuous string of points sweeping out the entire intersection curve. If the slicing plane is at the position L, then the intersection points I3 and I4 will collapse. Similarly, at the position H, the points I1 and I2 will collapse. These points are important because these are the positions where the intersection curve turns, i.e., these are the points where the tangent to the curve is parallel to one of the cone's (cylinder's) rulings. To find these points is important from the standpoint of accuracy and is done automatically as the above algorithm shows.

7.2.5. Including Break Points.

There are three kinds of break points that need to be included into the sequence of intersection points: (1) poles, (2) seam crossing points, and (3) double points. Poles appear when the entire boundary of the tensor product surface collapses into one point. Examples are the cone with one pole and the sphere with two poles. Poles are hard matters for numerical and topological reasons. They cause numerical problems because the partial derivatives vanish, and topological problems if the intersection curve passes through it.

The seam curve is a curve which is the image of two different parameter space lines. Examples are the cylinder, the cone, the torus, the sphere, and any other closed surface that is the map of a rectangular domain. For example, the cylinder's seam line is the image of the $u = 0$ as well as the $u = u$max parameter lines (assuming that it is ruled in the v direction). Therefore if the model space curve crosses the seam curve, the parameter space pre-image has to be split. To satisfy the requirement stated above, all three curves have to be split, i.e., the seam crossing points have to be included in the series of intersection points.

Double points appear in the case of tangential intersection. To make subsequent topological analysis simple, it was found to be useful to split the curve at the double point, e.g., a closed curve with a double point is considered as two closed curves without double points.

Fig. 7.6 illustrates how to include seam crossing points in the case of cylinder-sphere intersection. The seam crossing points are obtained by including a plane into the series of slicing planes that contains the seam line and hence results in intersection points that are actually break points. If there are more intersection points on one slicing plane than seam points, then they have to be mapped back to the parameter space of the surface to select those points whose pre-images are on the $u = 0$ or $u = u$max parameter lines (assuming that the cylinder is ruled in the v direction).

7.2.6. Handling Special Cases.

There are three cases that need to be handled carefully: (1) obtain the correct parameter values of seam crossing

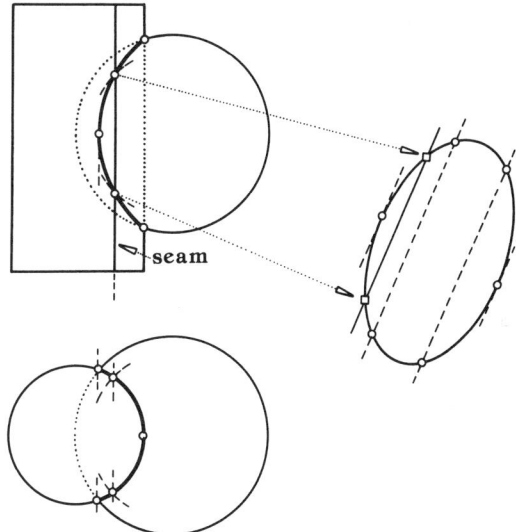

FIG. 7.6. *Including seam crossing points into the series of general intersection points.*

points, (2) compute the correct parameter values of poles, and (3) compute tangents at double points.

7.2.6.1. Pre-images of seam crossing points.

If an intersection curve crosses the seam curve of one of the surfaces, then its pre-image will break, that is, if **S** denotes this point, then **S** will be the beginning of one segment and the end of another. In other words, **S** will have two pre-images; one sits on the $u = 0$ line and the other on the $u = u$max line, assuming that the seam line is a v line. The question now is how to figure out which pre-image point belongs to which segment on the parameter domain. Assuming that a sufficiently dense point set is computed along the intersection curve, which is normally the case in SSI, the following simple process obtains correct parameter values:

```
get_seam_u-value(I[k-1],I[k],I[k+1],u[k])
{
  u[k-1]=pre-image(I[k-1]); u[k]=pre-image(I[k]);
  u[k+1]=pre-image(I[k+1]);
  if(I[k] is the beginning of the segment)
  {
    if(u[k+1]<u[k-1]) u[k]=0.0; else u[k]=umax;
  } else /* I[k] is the end of the segment */
  {
    if(u[k+1]<u[k-1]) u[k]=umax; else u[k]=0.0;
  }
}
```

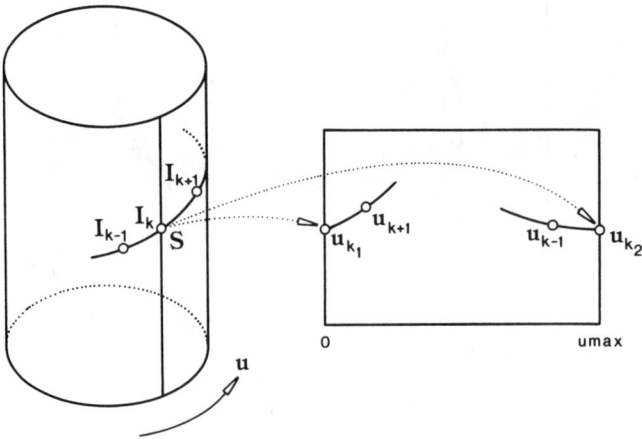

FIG. 7.7. *Computing correct u values of pre-images of seam crossing points.*

Figure 7.7 illustrates the process graphically.

7.2.6.2. Pre-images of poles.   Intersection curves passing through the pole have very interesting pre-images. While they are continuous curves in 3D, their pre-image curves will have discontinuities at the pre-images of the pole. Assume that the pole is the intersection point denoted by I_k and there is one more intersection point on each side of it, I_{k-1} and I_{k+1} (Fig. 7.8 depicts the situation using the sphere). Now, moving along the intersection curve from I_{k-1} to I_{k+1} and mapping back the curve segment result in two segments; one goes from u_{k-1} to u_{k_2}, and the other goes from u_{k_1} to u_{k+1}. The critical issue here is to compute the precise values of u_{k_1} and u_{k_2}. Using the sphere example in Fig. 7.8, these values are obtained as follows: each v isoparametric line is a semicircle running from the south pole to the north pole. If we consider the tangent at the pole, then the plane defined by the two poles and by the tangent will contain two isolines whose u_1 and u_2 values will be the parameters we are looking for. This can be seen as follows: each intersection point along with the two poles determines a semicircle isoline and a corresponding u value. Now, as the point approaches the pole, the corresponding u value will tend to u_1 or u_2. This method is applicable to the sphere only and has to be customized for other types of surfaces such as the cone.

7.2.6.3. Tangents at the double point.   Tangents at intersection points are obtained by taking the intersection of the tangent planes of the two surfaces (see Fig. 7.9(a)). Since at the double point these planes are parallel, this technique will fail to work. Obviously, the curve does have tangents at the double point; however, the above technique cannot be used to get them.

Fortunately, there is a very slick technique that is easy to implement and is applicable to any pair of surfaces. It is based on the *Dupin indicatrix* [18]. It is constructed as follows (see Fig. 7.9(b)): At an arbitrary point **P** of the

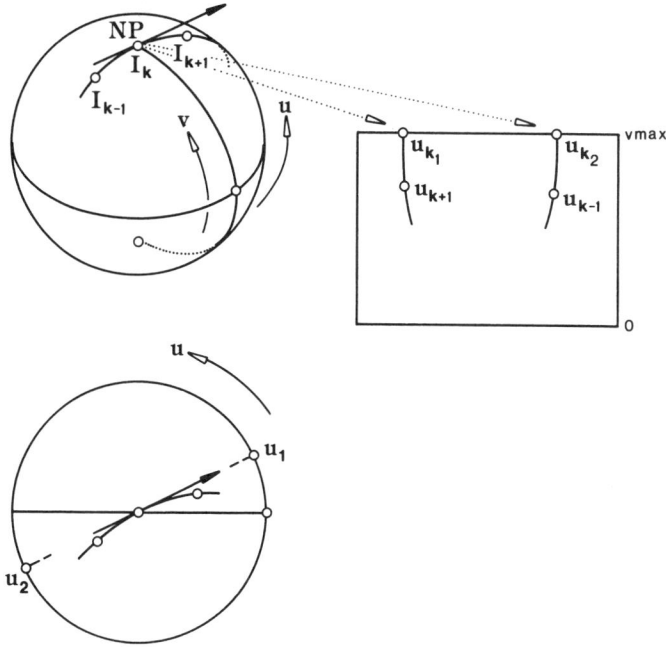

FIG. 7.8. *Computing pre-images of an intersection point sitting at the pole.*

intersection curve, take the tangent plane Π and the surface normal $\mathbf{m}$ of one of the surfaces (the two normals are parallel and hence it does not matter which one is chosen). Each tangent line $\mathbf{t}$ lying in Π and passing through $\mathbf{P}$ determines a plane perpendicular to Π and containing the surface normal $\mathbf{m}$. This plane is called the *normal section plane*. Obtain the intersection curve of the surface with the normal section plane and compute the curvature κ at $\mathbf{P}$. Let d be $1/\sqrt{|\kappa|}$ and measure d along $\mathbf{t}$ on both sides of $\mathbf{P}$. Now, if we rotate the normal section plane about the surface normal $\mathbf{m}$, then the locus of points obtained by measuring d on both sides of $\mathbf{P}$ is a *conic section* called the Dupin indicatrix. This indicatrix can be used to obtain the two tangents at the double point as follows: Since there are two surfaces, construct two Dupin indicatrix conic sections. These conic sections, being central symmetric, give four points of intersection which lie along diagonal lines passing through the center point $\mathbf{P}$. These diagonal lines will be the tangents we are looking for (Fig. 7.9(c) illustrates the situation for the case of cylinder-sphere intersection). To prove that the above construction is correct, first we quote a few theorems whose proofs can be found in books on classical differential geometry [19],[20] (refer to Fig. 7.9(d) for a better understanding).

THEOREM 7.2.1. *The curvature of a parametric curve is a geometric measure independent of the parameterization.*

The intersection curve is a curve on two surfaces and hence it can be parameterized in two different ways. The above theorem says that

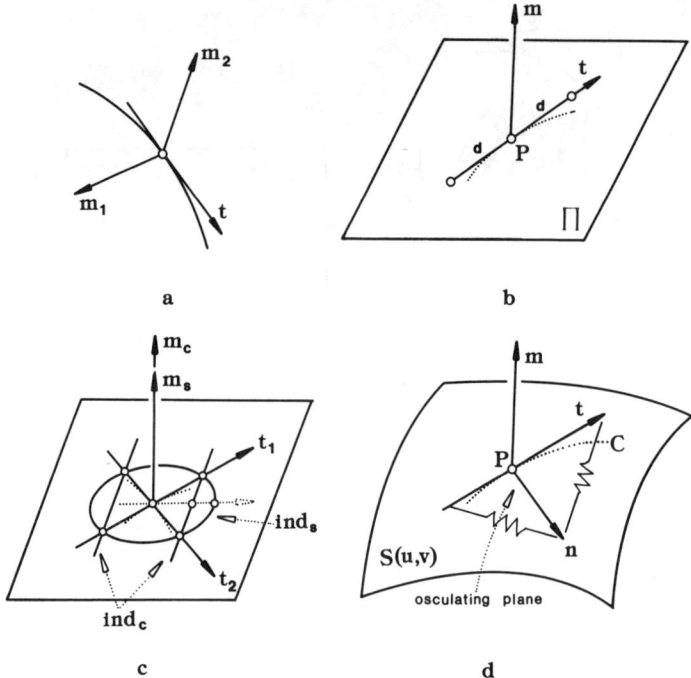

FIG. 7.9. *Computing tangents at the double point:* (a) *general method applicable at regular intersection points,* (b) *the definition of the Dupin indicatrix,* (c) *tangents computed at a double point via the Dupin indicatrix (cone-cylinder case), and* (d) *curve lying on a surface and associated geometric data required to compute the curvature at the point* **P**.

the curvature of the intersection curve is independent of the surface to be considered.

THEOREM 7.2.2. *At a given point* **P** *the curvature of a curve* C *lying on the surface* $S(u,v)$ *is given by*

$$\kappa = \frac{1}{\mathbf{m} \cdot \mathbf{n}} \frac{L\dot{u}^2 + 2M\dot{u}\dot{v} + N\dot{v}^2}{E\dot{u}^2 + 2F\dot{u}\dot{v} + G\dot{v}^2},$$

where **m** *is the surface normal at* **P**; **n** *is the normal to* C *at* **P**; E, F, *and* G *are the first fundamental forms; and* L, M, *and* N *are the second fundamental forms.*

This theorem says that the curvature of a surface curve is uniquely determined by the surface normal, by the curve normal and by the curve tangent (since $\mathbf{t} = S_u \dot{u} + S_v \dot{v}$).

COROLLARY 7.2.1. *If two surface curves meeting at* **P** *have the same osculating planes, then they have the same curvature as well.*

COROLLARY 7.2.2. *At a given point* **P** *the surface curve's curvature is equal to the curvature of the intersection curve obtained by intersecting the surface with the osculating plane of the surface curve.*

These corollaries say that when investigating the curvature of the intersection curve at a given point, it is sufficient to consider the intersections of the osculating plane with one of the surfaces.

THEOREM 7.2.3 (Meusnier). *If κ_n is the curvature of the normal section curve and κ is the curvature of another section curve obtained by intersecting the surface with a plane obtained by rotating the normal section plane about the tangent* **t**, *then*

$$\kappa_n = \kappa \cos \alpha,$$

where α is the angle of the rotation.

This theorem, due to Meusnier [21], says that the curvature of a planar intersection curve, e.g., the curve obtained by intersecting the surface with the osculating plane, can be obtained from the normal section curve's curvature.

COROLLARY 7.2.3. *At a given point* **P** *the curvature of the normal section curve uniquely determines the curvature of the intersection curve.*

Based on the above theorems and corollaries, the verification of the tangent construction is straightforward. The Dupin indicatrix gives the normal section curvatures which, on the other hand, determine the curvatures of surface curves passing through **P**. Let us consider a branch of the intersection curve only with tangent $\mathbf{t}_1$ at **P** and assume that this tangent does not pass through the intersection of the two Dupin indicatrices (Fig. 7.9(c)). Then the two Dupin indicatrices give two normal section curvatures, one for surface 1 and one for surface 2. These curvatures yield two curvatures of the intersection curve, which contradicts our first theorem. Therefore the tangents must pass through the intersections of the Dupin indicatrices.

7.2.7. Clipping.   Since in practical applications trimmed surfaces are intersected, the intersection curves that are the results of intersecting unbounded surfaces have to be clipped back to the bounded domain (classical descriptive geometric methods do not consider trimmed surfaces). Two kinds of clipping can be considered: (1) model space clipping and (2) parameter space clipping. Model space clipping means that only intersection points lying on both trimmed surfaces are computed. This works only for certain surfaces such as the cylinder and the sphere, where we can take advantage of the mutual symmetry of the two objects to discard points that lie outside the trimmed regions. Fig. 7.10 depicts a typical situation. The slicing plane is perpendicular to the cylinder's axis. The extreme positions of the series of slicing planes in the case of the cylinder are B_c and T_c, the planes of the bottom and the top circles. In the case of the sphere, these are the intersections of the contour circle of the sphere and the contour lines of the cylinder, denoted by B_s and T_s. Projecting these points onto a line parallel to the cylinder's axis and taking the intersection of the respective intervals yield the points BOT and TOP. Now, if a slicing plane passes through a point between BOT and TOP, then it will intersect the cylinder and the sphere in two circles whose intersection points will lie on both trimmed surfaces.

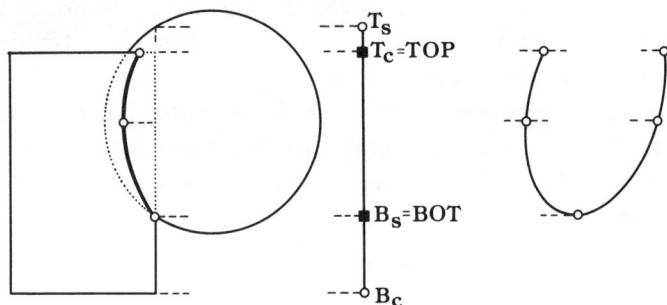

FIG. 7.10. *Clipping a model space intersection curve by specifying the extremal positions of the slicing planes.*

The above technique works only for a very few number of surfaces, and therefore parameter space clipping has to be applied in the rest of the cases. It involves the computation of the intersection curve of the *unbounded* surfaces and clipping away those segments that are outside the trimmed region. This is done by mapping each intersection point back into the two-parameter domains and checking whether the pre-image points lie inside the parameter rectangles. If they do, then we flag the intersection point as the *in* point, otherwise we flag it as the *out* point. To satisfy the requirements mentioned earlier, we need to flag each seam, pole, and double point as a *break* point. The actual clipping is done as follows:

```
clip_intersection( ...  )
{
  label each intersection point as "in", "out" or "break";
  start processing at a point labeled as "break" or "out";
  while( not all points are considered )
  {
    get next labeled point;
    if("break" to "break") output segment;
    if("break" to "out" || "out" to "break")
      { get the crossing point; output segment; }
    if("out" to "out") { get two crossing points; output
segment; }
    {
  }
}
```

Obtaining a crossing point means that we need to intersect one of the boundary curves of surface 1 with surface 2. In the case of quadric surfaces, this amounts to intersecting lines with quadrics or conic sections with quadrics. Line-quadric is not a problem; however, conic-quadric, which yields a degree 4 equation, is. Numerical analysts swear that this can be solved very easily; however, our practical experience with surface intersection showed that the coefficients

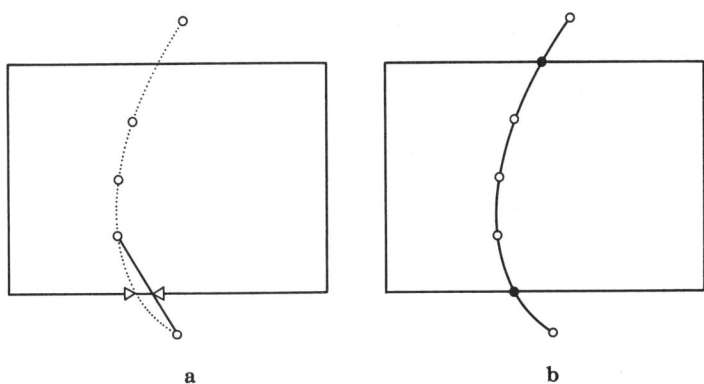

FIG. 7.11. *Parameter space clipping: (a) numerical technique that involves intersecting one of the boundaries with a high degree curve using a "good" guess value, and (b) approximate solution obtained by intersecting the boundary with a low degree curve fitted to the pre-images of intersection points.*

of the resulting equations can vary between $-\infty$ and $+\infty$ and therefore it is particularly difficult to set any kind of sensible tolerance [11]. One possibility is to approximate the crossing point as illustrated in Fig. 7.11(a), hoping that our favorite numerical method will provide a sufficiently accurate solution (many MCAE applications require high accuracy, e.g., 10^{-7}). However, a much easier approximate solution is as follows: Fit a low degree curve, e.g., a quadratic, through the *in* and one or two of the *out* points and intersect the boundary of the parameter space with the fitted curve (Fig. 7.11(b)).

7.2.8. Linking the Segments. There are three kinds of segments to be linked together: (1) conic segments, (2) clipped model space curves, and (3) unclipped model space curves. Conic segments are linked as follows (see Fig. 7.12):

```
link_conic( ...  )
{
  get all the break points and sort them;
  for( i... )
  {
    M=midpoint(segment[B[i],B[i+1]);
    m=pre-image(M);
    if(m is outside the parameter domain) continue;
    output_segment(B[i],B[i+1]);
  }
}
```

Linking clipped model space curves amounts to

- sorting the break points along the intersection curve, and

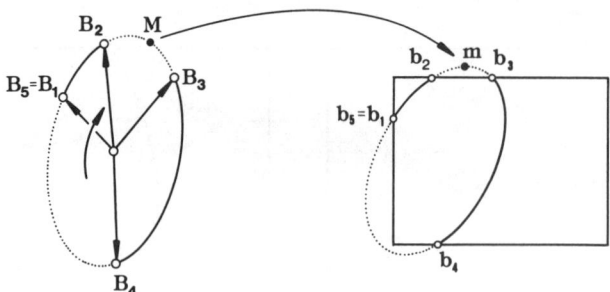

FIG. 7.12. *Linking conic intersection curves.*

- outputting segments between two break points watching for "jump." A jump occurs when the intersection curve consists of two open segments and the linker moves from one segment to the other.

Unclipped model space curve segments are linked by the parameter space clipper as discussed above.

7.2.9. Representing the Intersection Curves.

Basically there are two types of intersection curves to be represented: (1) conic segments and (2) general intersection curves given by a series of points. Since the input to the intersector was two NURBS surfaces and since the modeler we were working with could handle NURBS only, the intersection curve had to be represented in NURBS form.

We developed a general routine to represent any 3D conic arc in NURBS form using as convenient an input as possible [22] (the intersector is complicated enough and thus one desires that utility routines be as easy to use as possible). The algorithm takes as input the end points and the end tangents of the segment and one point anywhere along the curve, and outputs a quadratic NURBS curve that represents the arc precisely. Pre-images of conic segments are either conics or rational cubics or higher order curves depending on the surfaces intersected. Conics and cubics can be represented precisely, whereas higher order curves need to be approximated by computing a series of points (an example is a plane-torus intersection where the intersection curve can be two circles whose pre-images are high degree curves — an example is shown below).

The most important curves from the standpoint of Boolean operations are the parameter space curves. To represent them as precisely as possible, a large number of intersection points are computed. In order to overcome data explosion, we developed a fitting routine that approximates these intersection curves with quadratic curves up to a given user-defined tolerance [23]. The routine can approximate the points up to any given tolerance, and in case of zero tolerance it will degenerate to interpolation.

A special 3D fitting routine was also developed for data reduction purposes [24]. It works with rational cubics to allow the generation of twisted curves.

7.3. Computational Issues

The algorithms outlined in this paper have the following advantages:

- They are very stable and reliable as virtually no numerical methods are applied.
- All the intersection curves are found and special cases are handled accurately.
- Intersection points are subject to round-off errors only and the accuracy of the intersection curves depend on the accuracy of the curve fit employed.
- The algorithms are very fast. On a 1 MIPS machine, the intersection curves are found, computed, and output in less than a second.

There are, however, several drawbacks which should not be overlooked:

- Each surface pair needs a different routine.
- Each routine contains significant amount of code (1000–1500 lines C code).
- The topology of the intersection curve has to be established by the implementor.

Our experience showed that, despite the above disadvantages, the method presented herein was far superior to a general intersector [25] in speed, accuracy, as well as in reliability.

7.4. Examples

We show a few examples here to illustrate the power and usefulness of the method outlined in this paper. Figure 7.13 shows a trimmed cone and a trimmed cylinder intersecting in three curve segments, one closed and two open ones. S11 is the cone's label; S21 is the cylinder's label. Consequently the upper right part of the picture shows the parameter space curves in the cone's parameter space domain, and the lower right portion shows the pre-image of the intersection curves in the cylinder's parameter space domain. Notice that the closed model space curve is split into two segments labeled I3 and I4 because it crosses the cylinder's seam line.

Figure 7.14 illustrates the case when the general degree 4 intersection curve degenerates into two conic sections. Again, a trimmed cone and a trimmed cylinder are intersected resulting in two elliptical arcs. It is interesting to note that while the intersection is a conic section, its pre-images are rational cubic curves.

Plane-quadric intersections are depicted in Fig. 7.15 where a box is intersected with a cone. The right-hand side of the picture shows the parameter domain of the cone.

A double point intersection is illustrated in Fig. 7.16 showing a cylinder that penetrates the cone while touching it from inside.

A rather interesting intersection is shown in Fig. 7.17 where a cylinder penetrates through a cone in such a way that they share a ruling. Since this

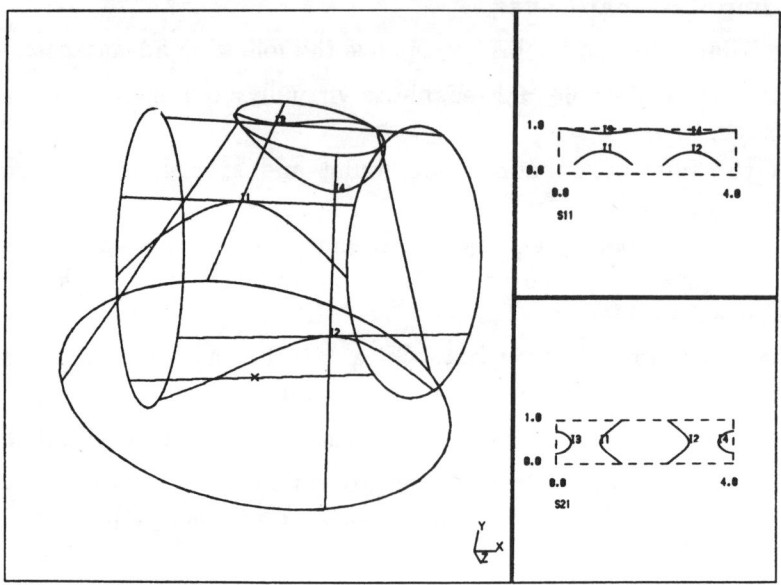

FIG. 7.13. *Intersection of trimmed cone and cylinder.*

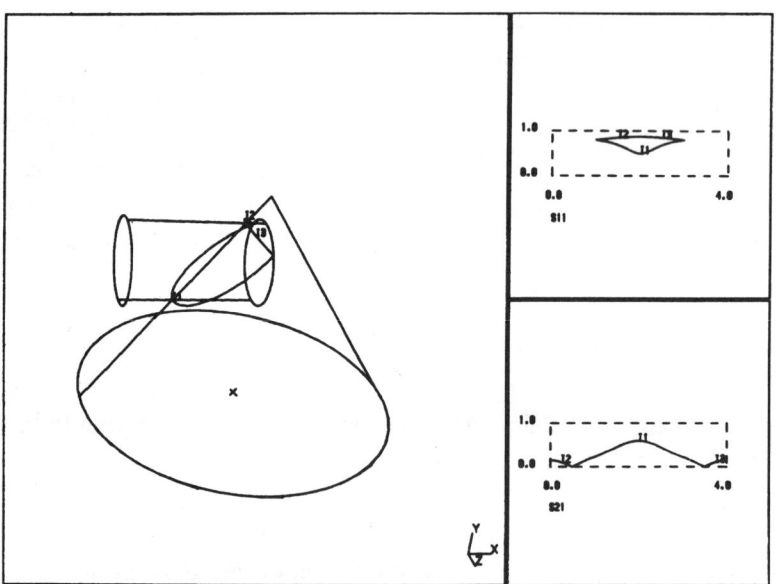

FIG. 7.14. *Cone-cylinder intersection resulting in two elliptical arcs.*

line is part of the intersection, the degree 4 curve will consist of a line and a degree 3 curve.

A line of revolution (trimmed cone patch) is intersected with a block in Fig. 7.18. Although this surface does not have the seam problem, it has problems with proper clipping. The right-hand side of the picture shows the parameter space of the trimmed cone patch.

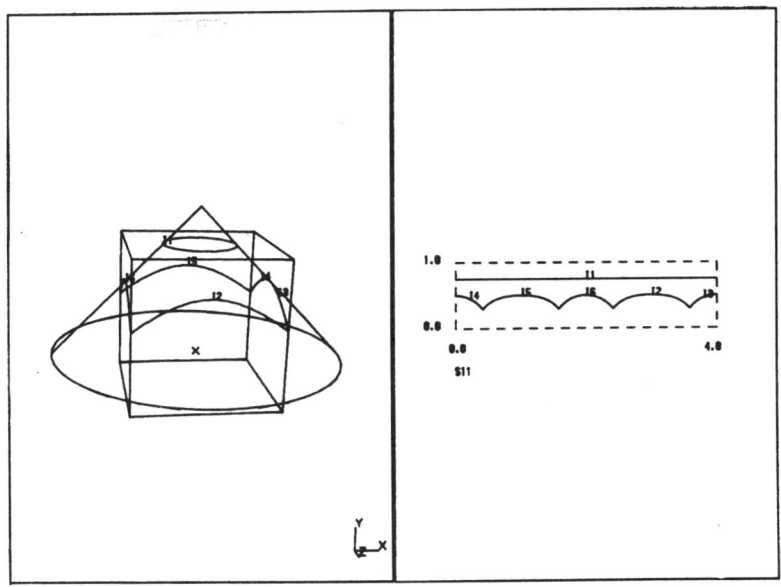

FIG. 7.15. *Box intersected with a cone.*

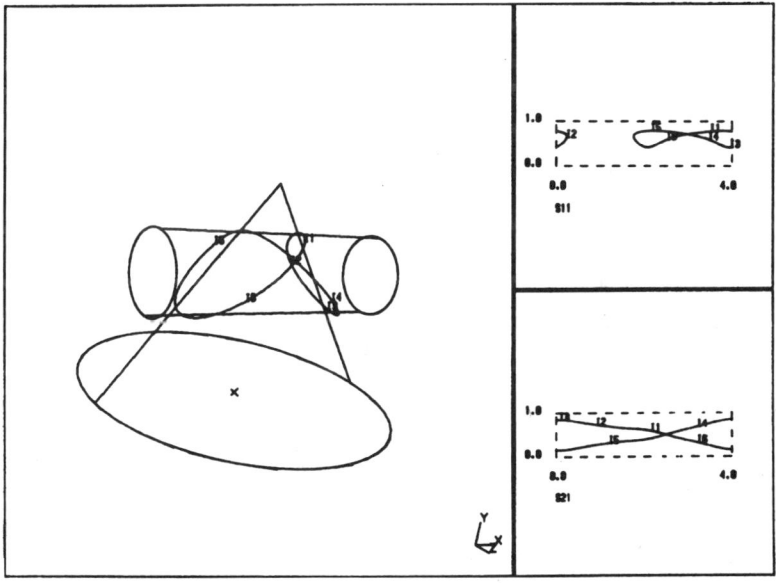

FIG. 7.16. *Cylinder penetrates cone while sharing a common tangent plane.*

A special case of the torus-plane intersection is illustrated in Fig. 7.19 where the bottom plane of the block intersects the torus at two hyperbolic points while being tangential to it at the same time. The intersection, as mentioned above, consists of two circles. The upper right side of the illustration shows the torus' parameter space curves (notice that the pre-images of the circles are

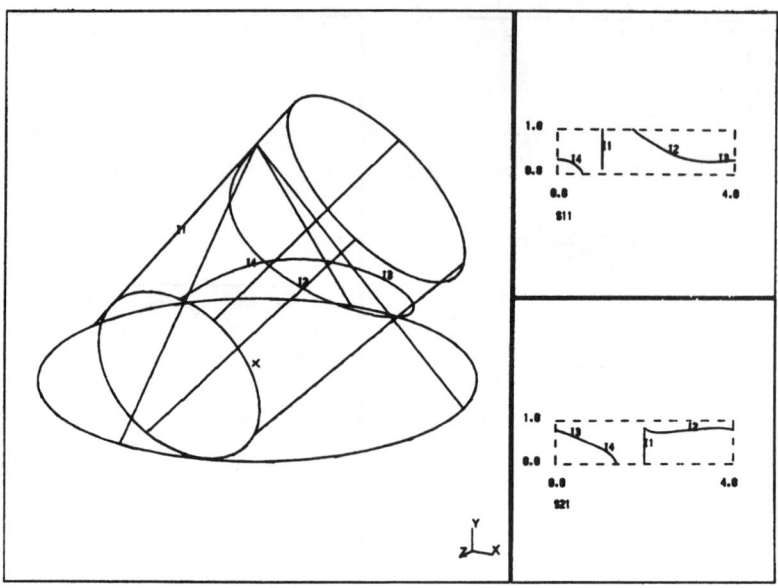

FIG. 7.17. *Cylinder-cone intersection sharing a common ruling.*

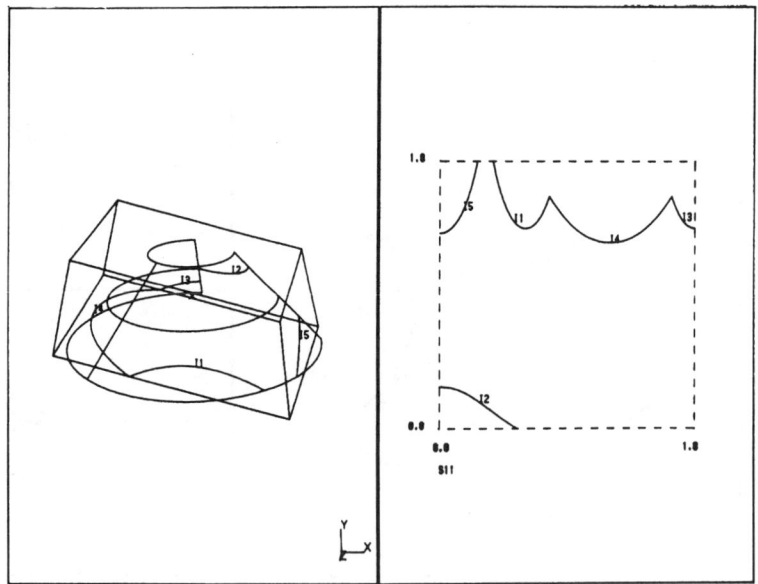

FIG. 7.18. *Line of revolution intersected with a box.*

high degree curves) and the lower right side shows the plane's parameter space curves.

A circular arc of revolution (trimmed toroidal patch) is intersected with a block in Fig. 7.20. Such surfaces appear very frequently in rounding and blending operations.

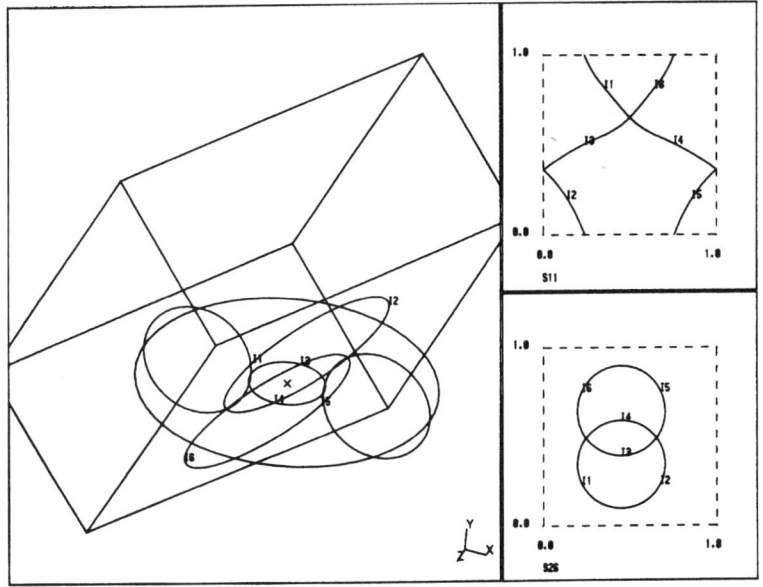

FIG. 7.19. *Bottom plane of box intersects torus while being tangential to it in two hyperbolic points.*

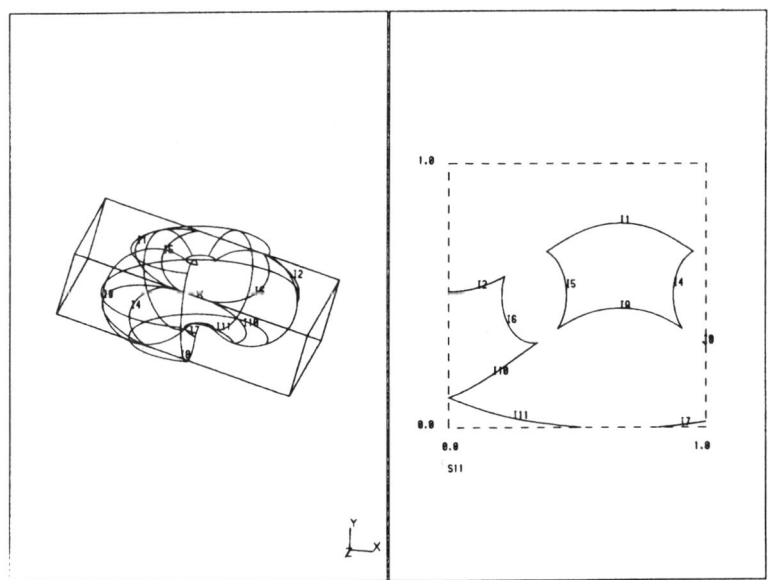

FIG. 7.20. *Box intersected with circular arc of revolution.*

Our final example is block-extruded surface intersection depicted in Fig. 7.21. The right-hand side of the picture shows the extruded surface's parameter space domain.

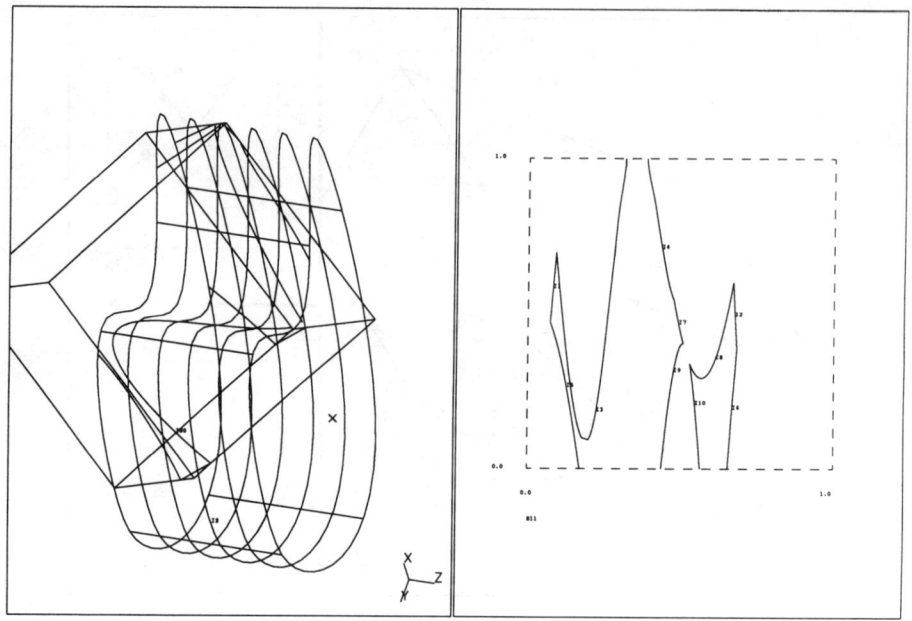

FIG. 7.21. *Box-extruded surface intersection.*

Acknowledgments

I would like to thank Wayne Tiller and Jin Chou for the many valuable discussions on surface-surface intersection. I am grateful to the reviewers for their valuable comments. I owe special thanks to the editor, Bob Barnhill, for the invitation to this volume that led to the writing of this paper. This research was supported in part by the Florida High Technology and Industry Council.

References

[1] G. Monge, *Géométrie descriptive*, Paris, France, 1798.

[2] K. Rohn and E. Papperitz, *Lehrbuch der Darstellenden Geometrie*, Leipzig, Germany, 1906.

[3] G. Loria, *Vorlesungen über Darstellenden Geometrie*, Leipzig, Germany, 1907.

[4] F. Hohenberg, *Konstruktive Geometrie für Techniker*, Wien, Austria, 1956.

[5] K. Strubecker, *Vorlesungen über Darstellenden Geometrie*, Göttingen, Germany, 1958.

[6] E. Müller and E. Kruppa, *Lehrbuch der Darstellenden Geometrie*, Wien, Austria, 1961.

[7] D. G. Hakala, R. C. Hillyard, B. E. Nourse, and P. J. Malraison, *Natural quadrics in mechanical design*, Autofact West, 1 (1980), pp. 363–378.

[8] J. Y. S. Luh and R. J. Krolak, *A mathematical model for mechanical part description*, Comm. CACM, 8 (1965), pp. 125–129.

[9] R. F. Sarraga, *Algebraic methods for intersection of quadric surfaces in GM-SOLID*, Computer Vision, Graphics, and Image Processing, 22 (1983), pp. 222–238.

[10] J. R. Miller, *Analysis of quadric-surface-based solid model*, IEEE Comput. Graphics Appl., 8 (1988), pp. 28–42.

[11] P. Linz, *A critique of numerical analysis*, Bull. Amer. Math. Soc., 19 (1988), pp. 407–416.

[12] L. Piegl and W. Tiller, *Curve and surface constructions using rational B-splines*, Comput. Aided Des., 19 (1987), pp. 485–498.

[13] L. Piegl, *Geometric method of intersecting natural quadrics represented in trimmed surface form*, Comput. Aided Des., 21 (1989), pp. 201–212.

[14] B. Kerekjarto, *Grundlagen der Geometrie: Projective Geometrie*, Budapest, Hungary, 1944.

[15] J. Z. Levin, *A parametric algorithm for drawing pictures of solid objects composed of quadric surfaces*, Comm. CACM, 19 (1976), pp. 555–563.

[16] _____, *Mathematical methods for determining the intersection of quadric surfaces*, Computer Graphics and Image Processing, 11 (1978), pp. 73–87.

[17] A. Yvone-Villarceau, *Extrait d'une note concernant un troisiéme systéme des sections qu'admet le tore circulaire*, Comptes rendus de l'Académie des Sciences, 27 (1848), p. 246.

[18] Ch. Dupin, *Développements de géométrie*, Paris, France, 1813.

[19] M. do Carmo, *Differential Geometry of Curves and Surfaces*, Prentice-Hall, Englewood Cliffs, NJ, 1976.

[20] H. Guggenheimer, *Differential Geometry*, Dover, New York, 1977.

[21] Ch. Meusnier, *Mémoire sur la courbure des surfaces*, Mém. des savants étrangers, 10 (1785), pp. 477–510.

[22] L. Piegl, *Algorithms for computing conic splines*, ASCE J. Comput. Civil Engrg., 4 (1990), pp. 180–198.

[23] _____, *A technique for smoothing scattered data with conic sections*, Comput. Industry, 9 (1987), pp. 223–237.

[24] J. Chou and L. Piegl, *Method for fitting cubic NURBS curves to 3-D data*, CSE 90-22, Department of Computer Science and Engineering, University of South Florida, Tampa, FL, January 1991.

[25] K.-P. Cheng, *Using plane vector fields to obtain all the intersection curves of two general surfaces*, in Theory and Practice of Geometric Modeling, W. Straßer and H.-P. Seidel, eds., Springer-Verlag, Berlin, New York, 1989.

Interrogation of Surface Intersections

Nicholas M. Patrikalakis

8.1. Introduction and Outline

Intersection provides the most basic relation of two surfaces, and, therefore, interrogation of surface intersections is a fundamental problem in geometric modeling of complex shapes in a computer environment. Such computation is needed primarily in the evaluation of the result of set operations on primitive volumes necessary in the creation of boundary representations of engineering artifacts. Such capability is useful in the design of complex objects, such as, for example, the internal subdivision and structural reinforcement of marine and aerospace vehicles, in numerous analysis tasks (such as finite element discretizations of three-dimensional solids), in feature recognition, and in simulation and control of manufacturing processes [10].

This paper is structured as follows. Section 8.2 provides a brief review of surface intersection computation techniques. The paper next focuses on methods, recently developed by the Design Laboratory, which promise to provide reliable processing of surface intersections in an efficient manner. Specifically, §8.3 focuses on the computation of intersections of algebraic (implicit polynomial) surfaces with piecewise rational, polynomial parametric surface patches (such as rational B-spline patches). Our method exploits the convex hull and stability properties of analytic representation of the algebraic curve of intersection in the Bernstein basis within the rectangular parametric domain of each rational polynomial element of the parametric patch. Section 8.4 focuses on methods allowing computation of intersections of two piecewise rational, polynomial parametric surface patches, such as two nonuniform rational B-splines. These are based on convex hull properties of rational B-splines and a combination of subdivision, vector field, topological, and differential equation techniques. Finally, §8.5 describes some outstanding problems in the area of surface intersections.

8.2. Recent Approaches to Intersection Problems

The fundamental issue in intersection problems is the efficient discovery and description of *all* features of the solution with high precision commensurate with the tasks required from the underlying geometric modeler [27]. Reliability of intersection algorithms is a basic prerequisite for their effective use in any geometric modeling system and is closely associated with the way features of the solution such as constrictions (near singular or singular situations), small loops, and partial surface overlap are handled. Reliability is a key property of automated solution procedures in a computer environment. If such property is not guaranteed, then the designer needs to manually ensure that the resulting approximations do not result in global inconsistencies of the geometric database. Such manual corrections are an extremely tedious and complex process. Global inconsistencies are very undesirable in a true, unified computer environment supporting design, analysis, and fabrication and, if present, may largely negate the potential advantages of full computerization expected from future systems. The solutions resulting from present state-of-the-art techniques are further complicated by imprecisions introduced by numerical errors present in all finite precision computations. These imprecisions may, by themselves, also introduce global inconsistencies of geometric data and further exacerbate the quality of available techniques. Surface intersection methods can be classified in four main categories: analytic, lattice evaluation, marching, and subdivision. Most of the methods have been developed in the context of polynomial surfaces, and, apart from the notable exception of some recent articles by Barnhill et al. [5], and Barnhill and Kersey [6], relatively little has been published on the intersection of more general classes of surfaces. A brief review of intersection techniques is attempted below.

Analytic methods rely on the derivation of a governing equation describing the intersection of two surfaces. For polynomial surfaces, the resulting equation is an algebraic curve which is an implicit polynomial in two variables. This equation, in principle, can be obtained either by elimination of one cartesian coordinate for the case of two implicit surfaces [39] or by elimination of three cartesian coordinates for the case of an implicit surface intersecting a rational polynomial parametric surface [13]. The intersection between two algebraic surfaces can be handled as an algebraic–rational polynomial parametric surface intersection when a rational polynomial parameterization of one algebraic surface is possible or another polynomial parametric surface which contains the curve of intersection can be found. The second approach has been used in the intersection of two quadrics [24], [38] where a third parametric polynomial-ruled surface is found which also contains the intersection between the quadrics. The intersection between two rational polynomial parametric surfaces can be handled, in theory, by obtaining an algebraic (implicit polynomial) representation for one of the surfaces [17]. The relatively high degree of this algebraic representation and the subsequent substitution of the second rational polynomial parametric surface into this high degree equation leads to an algebraic curve of even higher degree, making this approach unsatisfactory from the numerical stability and efficiency points of view.

Once the equation describing the algebraic intersection curve between low degree algebraics and rational polynomial parametric patches is obtained as above, it must be traced. For special cases, the resulting implicit equations can be solved in terms of explicit expressions involving radicals [24] or in terms of rational parametric polynomials for algebraic curves with known singularities and whose genus is zero [1]. Once the range of the independent variable in such cases is determined, the above explicit equations can be used to trace the intersection curve. The local extrema and singular points of the curve can also be used to our advantage [38]. However, for general cases explicit representation of algebraic curves in terms of elementary functions are impossible [41]. The class of algebraic curves with integer coefficients can be analyzed using the cylindrical algebraic decomposition algorithm [3]. This method, as implemented in rational arithmetic, although providing a guarantee that the solution is topologically reliable, is impractical because of its very large memory requirements and poor efficiency.

Lattice evaluation methods reduce the dimensionality of surface intersection problems by computing intersections of a number of isoparametric curves of one surface with the other surface followed by the connection of the resulting discrete intersection points to form different solution branches [48], [37]. For intersections of parametric patches, the method reduces to the solution of a large number of independent systems of three nonlinear equations in three unknowns. The reduction of problem dimensionality in lattice methods involves an initial choice of grid resolution, which, in turn, may lead the method to miss important features of the solution, such as small loops and isolated points which reflect near tangency or tangency of intersecting surfaces. The connection of discrete solution points to form intersection curve branches requires determining adjacency on the basis of minimum mutual distance which may lead to incorrect connectivity, particularly at singular points or for near singular situations. Derivative information may be employed to enhance the reliability of the method in such cases.

Marching methods involve generation of sequences of points of an intersection curve branch by stepping from a given point on the required curve in a direction prescribed by the local differential geometry. However, such methods are by themselves incomplete in that they require *starting points* for every branch of the solution. Starting points are usually obtained using lattice and subdivision methods [5]. They also require a variable stepping size appropriate for the local length scales of the problem. Incorrect step size may lead to erroneous connectivity of solution branches or even to endless looping in the presence of closely spaced features [16]. Reliability of marching as well as lattice evaluation methods can be substantially improved by the determination of all *border, turning, and singular* points (collectively referred to as *significant points*) of the curve [38], [13]. Curvature analysis or power series expansions about each point of the solution to control the step size along the tangent also improve the reliability of marching methods [11]. For algebraic curves, the desingularization method based on birational transformations [4] allows marching through singularities. Knowledge of all significant points within a domain and the multiplicity of singular points also provides an independent count of the number of monotonic branches between significant points. This count is usually able to confirm the number of branches obtained using lattice and marching methods and provides added confidence in the solution [13].

Subdivision methods involve recursive decomposition of the problem into simpler, similar problems until a level of simplicity is reached, which allows a simple direct solution (e.g., plane/plane intersection). This is followed by a connection phase of the individual solutions to form the complete solution. Initially conceived in the context of intersections of polynomial parametric surfaces [23], [47], [8], they can be extended to the computation of algebraic/ rational polynomial parametric and algebraic/algebraic surface intersections [16], [39], [30]. Section 8.3 of this paper is devoted to our work on intersections of algebraic and rational polynomial parametric surfaces. Subdivision techniques do not require starting points as marching methods, an important advantage from the reliability point of view. Many elements of subdivision techniques are also parallelizable, which is an important advantage for future large-scale, real-time applications. General nonuniform subdivision allows selective refinement of the solution providing the basis for an adaptive intersection technique. Subdivision methods have been used in diverse ways to address intersection problems [34], [46]. A disadvantage of subdivision techniques is that, in actual implementations with finite subdivision steps, correct connectivity of solution branches in the vicinity of singular or near-singular points is difficult to guarantee; small loops may be missed or extraneous loops may be present in the approximation of the solution. Furthermore, if subdivision methods are used for high precision, they lead to data proliferation and are consequently slow.

As can be seen from the above review, common problems of state-of-the-art techniques include the difficulty in handling singularities, surface overlap, and efficiently identifying closely spaced features and small loops. Recently, there is strong interest in developing techniques to detect the existence of loops in an intersection curve [45], [30], [35], [36], [9], [43], [42], [31], [19]–[22]. The issue of loop detection and our work in this area is addressed in §§8.3 and 8.4 of this paper.

8.3. Algebraic Surface and Rational Polynomial Surface Patch Intersection

8.3.1. Formulation. The representation of planar algebraic curves within a rectangular domain in the Bernstein basis coupled with a priori computation of the significant points of the curve is employed in recent work by the Design Laboratory [30], [35], [36], [31], [22] to compute intersections of algebraic (implicit polynomial) surfaces and rational polynomial parametric surface patches conveniently expressed in the Bernstein basis (rational Bézier patches defined over rectangular parametric domains). Our method combines the advantageous convex hull and stability features of representation of the algebraic intersection curve in the Bernstein basis; the a priori computation of border, turning, and singular points *(significant points)*; and adaptive subdivision techniques to provide the basis for a reliable and efficient solution procedure. Using the splitting of a piecewise rational, polynomial parametric surface patch in its rational polynomial subpatches, interrogation of intersections of an algebraic surface with a rational B-spline surface patch can be reduced to the above method. Using elimination techniques, our method could be also extended to handle intersections of two algebraic surfaces for which rational polynomial parameterizations are not available.

Our research attempts to address some common problems in presently available algorithms such as those caused by surface overlap, singularities, or near singularities (constrictions) and the presence of small isolated features in the intersection curve such as small loops. Surface overlap is difficult to handle, particularly by methods not relying on an analytic representation of the intersection in terms of a single governing equation. Most marching algorithms have difficulty in handling singularities or closely spaced features.

Recent work in the desingularization of algebraic curves, coupled with singular point computation during the marching process and in the use of power series expansions about a point, has improved these techniques but a common problem is the computation of one point on every segment of the curve in a nonexhaustive and efficient manner. Lattice evaluation methods may miss small features of the curve because of the need to choose a grid size a priori. Published methods from this class also do not have efficient means of resolving singularities in the curve. Pure subdivision methods, although convergent in the limit, lead to data proliferation and are inefficient in accurately solving intersection problems, particularly involving singular points and small loops. The computation of the analytic representation of the curve provides the capability to transform the problem at hand to the intersection of an auxiliary, integral Bézier surface patch (*control surface*) and an (auxiliary) *control plane* and is a key feature of the method. This transformation allows the use of subdivision methods in the solution of algebraic surface-rational polynomial patch or algebraic surface-algebraic surface intersection problems and provides an efficient method of obtaining points on every segment of the curve.

In our work, we first investigated symbolic substitution of the rational polynomial patch equation into the algebraic (implicit polynomial) surface equation [30], [31]. The resulting expressions were next expanded to provide explicit expressions for each coefficient of the algebraic curve (in the Bernstein basis) in terms of primary data. If the primary data for the patch and the algebraic curve are both expressed in the Bernstein basis, the resulting expressions for the coefficients turn out to be compact. The availability of explicit expressions for each of the coefficients also allows the use of sophisticated summation schemes for large-scale series [35], which improve the accuracy in the coefficients of the algebraic curve in comparison to direct techniques. This method of computation of the coefficients of algebraic curves was implemented for cases involving intersections of planes and quadrics with up to rational bicubic surface patches [35]. However, for higher degree intersections such symbolic computation becomes prohibitively expensive in terms of space and processing time requirements. To avoid this undesirable feature of the method, we have developed a numerical evaluation procedure of the coefficients of the algebraic curve of intersection employing the Bernstein basis throughout the computation [22]. In this manner, we avoid intermediate use of the monomial (power) basis [14] and save some unnecessary intermediate operations. Our method is based on, possibly repeated, multiplication of Bernstein polynomials [46] and summation of appropriate coefficients. The technique we developed allows treatment of intersections of arbitrary degree algebraic surfaces with rational biquadratic and bicubic Bézier patches and has been successfully tested with up to degree 4 algebraic surfaces.

8.3.2. Significant Point Computation. The computation and use of border, turning, and singular points of the curve allows subdivision and faceting techniques to provide approximations of the curve with correct connectivity in an efficient fashion. It was found that border points could be isolated reliably using subdivision and the variation diminishing property of the Bernstein basis and computed efficiently using a modified Newton method.

Two methods of identifying turning and singular points were attempted. The first method involved the solution of univariate polynomials obtained by elimination techniques of algebraic geometry. It has been observed that the accuracy and efficiency obtainable by solution of such equations may be below the requirements of geometric modeling applications and, hence, this method is limited to very low degree cases [35], [36]. The second method involved the use of a relatively coarse linear approximation of the curve obtained by subdivision and faceting of the control polyhedron to generate initial approximations for turning and singular points to be used with minimization and Newton techniques for solving systems of nonlinear equations. In our numerical experiments using a very coarse subdivision of the polyhedron, the above procedure was in a position to provide *all* turning and singular points of a variety of algebraic curves chosen for their complexity and diversity [35]. However, no guarantee exists that in general at least one starting point exists for every turning point nor that the iterative technique will converge to the appropriate turning point. To enhance reliability, a theoretical verification of the success of the above process to determine all turning and singular points is made using the following additional computation [36], [31]. The control surface is first partitioned using all *available* border, turning, and singular points. Next, each subpatch is examined for possible intersection with the control plane, using the signs of coefficients of the algebraic curve in the Bernstein basis. This test eliminates nonintersecting subpatches. Next, the subpatches with possible intersection are examined for existence of any turning/singular points away from the corners. This can be performed using the first two partial derivative surface control points for each subpatch. Using the convex hull property, such polyhedra can verify the absence of turning/singular points not at corners of the subpatches (all control points except corner points are on one side of the control plane). In case of failure, the initial starting point computation and the minimization Newton technique is invoked within the smaller subdomain, and the above process of splitting at new turning/ singular points and checking via the derivative patch is repeated until no more turning points can be detected. The above method to obtain starting points and compute turning and singular points is very useful because it does not depend on external information to initiate the procedure. In addition, unlike pure subdivision methods, it practically allows accurate computation of turning/singular points with very coarse initial subdivision and quadratically convergent Newton-like iteration. Our method to identify turning and singular points, in principle, could be combined with marching methods because it provides all the necessary starting points on every segment of the curve and the singularities of the curve. These methods show promise as they do not involve elimination techniques and the solution of high degree polynomials and, as a result, provide improved accuracy on these significant points. It must also be noted that this method of computing turning and singular points is not as seriously restricted by degree as elimination methods since only evaluations of relatively low degree representations are required for the search solution procedures.

8.3.3. Intersection Curve Tracing and Concluding Remarks. The tracing of the curve may be accomplished by first splitting the control Bézier surface patch at all parametric lines passing through significant points leading to a matrix of smaller subpatches [35], [36], [31]. Each subpatch intersection, if not empty, contains monotonic curve segments (possibly more than one) that extend from border to border of the subpatch and are simpler to trace using subdivision methods. More recently, a way to partition the intersection domain in a smaller number of subdomains compared to those used above and to still ensure that intersection segments are monotonic and significant points are on a corner of one of the subpatches has been developed [22]. A *tree structure* is required to achieve this more economic partitioning of the domain. This method relies on sequential rather than simultaneous partition of the domain. For each significant point on the border of a subdomain, the domain is split into two pieces using a line orthogonal to that border. For each internal significant point, the domain is partitioned into four pieces. Thus, each of the significant points becomes a node in the tree with two or four leaves. When there are no remaining significant points in a subdomain, the partitioning stops. At the end of partitioning, the final leaves of the tree correspond to subdomains potentially containing monotonic intersection segments, which require tracing. The above splitting processes are important extensions to standard subdivision and faceting algorithms and provide subdivision-based intersection solutions with the potential for efficiently extracting the true connectivity of the curve.

Approximate tracing of curve segments in a subdomain may be based on a faceted approximation of the subpatch control polyhedron and intersection of this faceted approximation with the control plane. This approximation may be later refined using an efficient Newton technique. A tracing algorithm may use a triangulation of the control polyhedron of the subpatch to compute the intersection segments by intersecting triangular facets with the control plane [35]. An important problem with any triangulation algorithm is that for closely spaced intersection features (constriction), the direction of triangulation (of the two possible) affects the connectivity of the approximation. Local subdivision is required to resolve this type of problem [37], [36]. Another tracing scheme, not involving triangulation, which we recently developed is based on intersection of the edges of the quadrilaterals of the control polyhedron with the control plane [22]. The intersection information for the edges is assigned to each of the quadrilaterals of the polyhedron to assist in the connection phase of the intersection segments to form solution branches. The type of intersection in each quadrilateral is determined by the number of edges of the quadrilateral that have an intersection and the number of vertices that lie on the control plane. In the majority of cases, there is a single intersection segment per quadrilateral and connectivity is easy to determine. The quadrilaterals with intersections in all edges suggest the existence of a closely spaced feature (constriction). Whenever such a situation is detected in the interior of the control polyhedron, knots are added in the interior of the quadrilateral with intersections in all four edges, until all these situations are resolved and correct connection can be achieved. This is always possible because all singular points are found a priori and used to subdivide the intersection domain. The next step is the connection of linear intersection segments to form solution branches. These segments are connected into complete intersection branches by the use of control point indices to provide direct, unambiguous matching of adjacent linear segments [22]. Thus the tracing scheme provides connected segments between significant points of the curve, which may then be further connected together based on the application. Newton-based correction methods are further employed to efficiently enhance the precision of each approximate intersection point and chordal approximation to the curve. This addition is important in reducing the space and time requirements of subdivision methods when used to obtain the high precision required for intersections in geometric modeling applications. Newton method-based corrections, although efficient in space and time, do necessitate topological consistency verification within each subpatch (e.g., the creation of an artificial self-intersection within a subpatch). This can be performed using the final piecewise linear approximation to

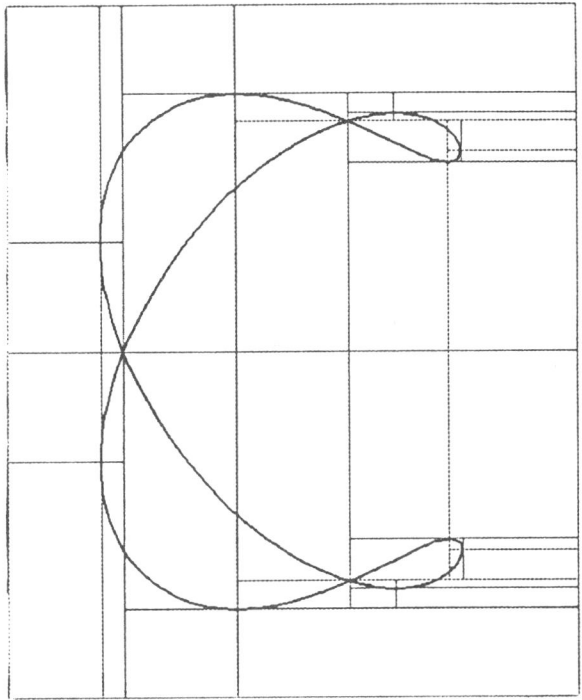

FIG. 8.1. *A degree 4 algebraic curve with three self-intersections* [22].

the curve. Alternatively, once all significant points are available, marching methods employing an adaptive stepping strategy may be employed.

A variety of complex algebraic curves and actual intersection problems with diverse features were solved successfully using the technique described above. Our method of tracing algebraic curves is illustrated in Fig. 8.1. The application of the method in the solution of intersection problems between algebraics (such as quadrics and tori) and piecewise rational polynomial surface patches (such as rational biquadratic and bicubic B-splines) is illustrated in Figs. 8.2 and 8.3.

A consistent effort has been made to carry out all evaluations and solutions for the roots of polynomials using the Bernstein basis without any intermediate conversions to the monomial basis to exploit the relative numerical stability of the Bernstein basis. The numerical reliability of the methods developed depends on the accuracy with which significant points are computed. Border and turning points are computed with very good accuracy. As expected, singular points are computed to a lesser accuracy which varies with the type and order of singularity. Cusps were found to be most difficult to compute accurately. These inaccuracies in the computation caused by imprecise floating point arithmetic should be reflected in the way the small constant parameters of the method are chosen and procedures to choose such constants in a dynamic manner, based on numerical error estimates, ought be explored. Because of the

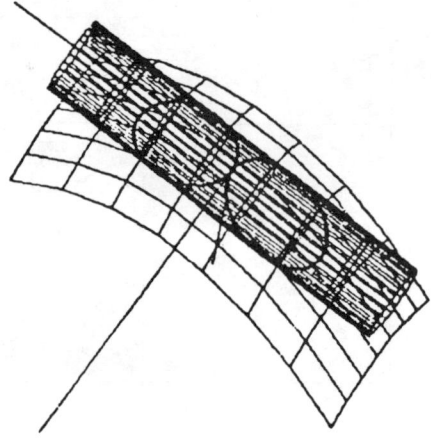

FIG. 8.2. *Cylinder and elliptic paraboloid intersection with a cusp* [31].

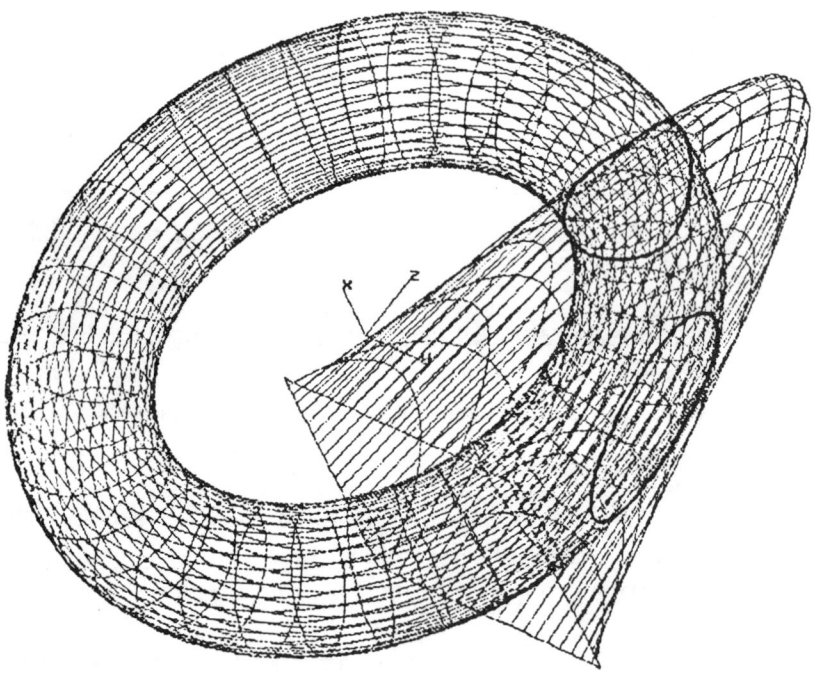

FIG. 8.3. *Torus and bicubic patch intersection with two loops* [22].

inherent ill-conditioned nature of singularities, it may be required to compute singular points in extended precision.

Both the computation of significant points using the numerical solution methods and the tracing of the intersection curve are inherently parallel processes. This can be seen in at least two stages of our intersection method: (i) at the level of splitting the rational piecewise polynomial patch into its rational polynomial components, and (ii) in the subdivision of the control polyhedron and the intersection of each facet of the control polyhedron of each subpatch with the control plane. The parallel nature of the method could become important in future large-scale, real-time applications.

The methods used in our work depend to a large extent on the availability of an explicit polynomial representation of the intersection curve. Our method has been tested successfully in the intersection of algebraic surfaces up to degree 4 and rational polynomial parametric surfaces up to degree 3 in each parameter, for which a representation of the intersection can be obtained both accurately and efficiently. Extension of the method to the intersection of other low order algebraics, possessing an easily derivable implicit representation (such as surfaces of revolution with planar, rational polynomial, low degree profile curves, as well as conical or cylindrical ruled surfaces with low degree rational polynomial surfaces) is direct. The above representation of the intersection curve required by the method described above, although theoretically available for all intersections between rational polynomial surfaces, is impractical to obtain in some important cases. For example, for intersections between rational biquadratic and bicubic patches, such a representation is very difficult to compute efficiently and accurately [17], and alternate methods of obtaining the intersection were developed for such important cases, as outlined in §8.4.

An alternate method would be to study modeling sculptured shape with surface representations that would lead to algebraic intersection curves of lesser degree compared to the intersections of the widely used rational biquadratics and bicubics. Modeling with low order, piecewise algebraic surface patches using the barycentric Bernstein representation within tetrahedra or the B-spline representation within rectangular boxes provides such examples [40], [29]. The advantages and limitations of these surface representation methods are discussed in an earlier paper [27].

8.4. Intersection of Rational Polynomial Surface Patches

8.4.1. Formulation. Our research on the problem of intersecting two rational polynomial surface patches defined over rectangular parameter spaces and expressed in the Bernstein basis is summarized in this section [19]–[21]. Such surfaces (usually referred to as rational Bézier patches) arise, for example, by splitting rational B-spline surface patches into their rational polynomial subpatches. The present general problem, although less frequent than the intersection problem addressed in §8.3, is nonetheless necessary in geometric modelers supporting free-form or sculptured surfaces. Rational B-splines are usually employed as a canonical geometric representation for creation, interrogation, storage, and exchange of data in or between such modelers [10].

The above intersection problem can be viewed as the solution of a system of three nonlinear relations, equating the three cartesian coordinates of the two surfaces, $\vec{F}(u,v,s,t) = \vec{r}(u,v) - \vec{q}(s,t) = 0$, where $\vec{r}(u,v)$ and $\vec{q}(s,t)$ are the two surface patch equations involved. Alternatively, the intersection set can be viewed as the set of points on the two surfaces with zero distance. For this purpose, the oriented distance function ϕ between an arbitrary point on $\vec{r}(u,v)$ and the surface $\vec{q}(s,t)$ may be introduced:

$$(8.1) \qquad \phi(u,v) = \vec{n}_2[\vec{Q}(\vec{r}(u,v))] \bullet \{\vec{r}(u,v) - \vec{Q}(\vec{r}(u,v))\},$$

where $\vec{Q}(\vec{r}(u,v))$ is a point on surface $\vec{q}(s,t)$ which is the unique nearest point to the point $\vec{r}(u,v)$, and $\vec{n}_2$ is the unit normal vector on surface $\vec{q}(s,t)$ at the point $\vec{Q}(\vec{r}(u,v))$. Note that $|\phi(u,v)|$ is the actual Euclidean distance of point $\vec{r}(u,v)$ from point $\vec{Q}(\vec{r}(u,v))$ on surface $\vec{q}(s,t)$, provided the vectors $\vec{n}_2\{\vec{Q}(\vec{r}(u,v))\}$ and $\{\vec{r}(u,v) - \vec{Q}(\vec{r}(u,v))\}$ are collinear (*collinearity condition*). In this case, point $\vec{Q}(\vec{r}(u,v))$ is the orthogonal projection of point $\vec{r}(u,v)$ on surface $\vec{q}(s,t)$ and the first- and second-order derivatives of ϕ (referred to as a proper oriented distance function) may be computed *explicitly*. The intersection set between surfaces $\vec{r}(u,v)$ and $\vec{q}(s,t)$ is equivalent to the *zero set* of function ϕ, i.e., the set of points satisfying the implicit equation $\phi(u,v) = 0$, provided that $\phi(u,v)$ is proper.

There are special cases in which $\nabla\phi$ is not well defined. First, there may be more than one point $\vec{Q}(\vec{r}(u,v))$ on the surface $\vec{q}(s,t)$ nearest to point $\vec{r}(u,v)$. This occurs when $\vec{r}(u,v)$ belongs to the cut locus [49] of $\vec{q}(s,t)$. The cut locus is a generalization of the medial axis concept [7], [28]. It should be noted that for C^1 rational spline surfaces of at least order 3 which are regular [12], the cut locus of $\vec{q}(s,t)$ stays away from $\vec{q}(s,t)$ and, therefore, all these points where $\nabla\phi$ is not well defined are away from the intersection set. Second, when the surface $\vec{q}(s,t)$ has a finite boundary, the nearest point $\vec{Q}(\vec{r}(u,v))$ may lie on the border and the collinearity condition may not be satisfied. These types of points are always away from the intersection except from border intersection points, which are, however, explicitly computed by our methods. In practice, we found that the presence of these points does not affect the reliability of our algorithms.

A crucial element of the intersection problem is the identification of *all* connected intersection components. The properties of ϕ can assist in the identification of intersection components. A closed loop in the intersection set of the two surfaces corresponds to a closed level curve of ϕ in the parameter domain, i.e., $\phi(u,v) = c = 0$. If the domain D enclosed by such a curve is simply connected and ϕ is properly defined at all points in D, then that domain contains an *extremum* of $\phi(u,v)$, and consequently a *critical point* of ϕ. At a critical point $\nabla\phi = 0$. Therefore, the critical set of ϕ is directly related to the topology of the zero set of ϕ. If we could describe the structure of the critical set of ϕ, including ill-defined regions, then we would have sufficient information to identify all components of the intersection. Topological tools based on the *rotation number* of the vector field $\nabla\phi$ along a closed curve on the (u, v) parameter space are important to the solution of the intersection problem formulated above (see §8.4.2). The appropriate background theory may be found in [18], [2]. When the collinearity condition is valid, $\nabla\phi$ defined in the u, v parameter space assumes the following form [19]

(8.2) $\nabla \phi(u, v) =$
$$\{\vec{n}_2[\vec{Q}(\vec{r}(u,v))] \bullet \vec{r}_u(u,v), \vec{n}_2[\vec{Q}(\vec{r}(u,v))] \bullet \vec{r}_v(u,v)\}^T,$$

where subscripts u, v denote partial derivatives and superscript T denotes transpose. This field is also suggested by [9] as useful in computing the intersection of two parametric surfaces and the vector field used by [25] to compute singular points of the intersection using Newton iteration. A detailed critique of these two earlier references can be found in [21]. The set $\phi = 0$ provides an alternate *implicit representation* of the intersection which is analogous to the algebraic curve of intersection defined in the parameter domain of one of the patches as described in §8.3.

8.4.2. Significant Point Computation.

In order to identify all connected components of an intersection curve, a set of important points on the intersection curve (*significant points*) can be defined. As seen in §8.3.2, such a set includes *border, turning, and singular points* of the intersection and provides at least one point on any connected intersection segment and identifies all singularities. An alternate and more convenient set of such points, discussed in this section, sufficient to discover all connected components of the intersection, includes *border and collinear normal points* between two surfaces. Collinear normal points provide points inside all intersection loops and all singular points.

Border points. These are points of the intersection at which at least one of the parametric variables u, v, s, t takes a value equal to the border of the u-v or s-t parametric domain. The points with parametric values equal to the border of the s-t parametric domain are also called *termination points*. To compute border points, a piecewise rational polynomial curve to piecewise rational polynomial surface intersection capability is required. A combination of subdivision and numerical techniques has been used in our work to compute these points with high accuracy.

Collinear normal points. These are points on the two parametric surfaces at which the normal vectors are collinear. These points do not necessarily lie on the intersection, but if they do, they are also singular points. An alternative interpretation of collinear normal points can be obtained using ϕ. Most of the collinear normal points are *critical* points of ϕ, where $\nabla \phi = 0$, with ϕ not necessarily 0 there. If ϕ is proper at a point, then $\nabla \phi = 0$ at that point implies the presence of a collinear normal point pair there. There are some collinear normal points between two surfaces which are not critical points of ϕ, as they are not minimum distance point pairs. In this paper we are interested in computing the collinear normal points between two surfaces, which are critical points of ϕ. In intersection problems, collinear normal points and critical points of ϕ may be used to signify the same concept, because after some subdivision, necessary in surface intersection, the extraneous collinear normal points become irrelevant.

The importance of collinear normal points in detecting the existence of closed intersection loops in surface intersections has been well established during the last decade [46]. A process for verifying the absence of collinear normal vectors using an interval type of analysis was suggested in [42], [44]. The importance of computing collinear normal points and not just determining their absence and a linearly convergent subdivision method to detect these points are also discussed in [42], [44]. Recently, Newton techniques to compute collinear normal points which lie on the intersection, (i.e., singular points) were developed in [25]. This method does not provide, however, a technique to automatically determine initial approximations for all these points. In our work we have developed an effective method to compute initial approximations to collinear normal points. Direct numerical techniques (minimization and Newton methods) are subsequently used for the computation of these points. Such methods depend only on evaluations of the surface positions and their partial derivatives.

The theory of plane vector fields [18] may be used to determine these initial approximations. The vector field of interest is $\nabla\phi$ given in (8.3). The *rotation number* of $V = \nabla\phi$ along a closed curve γ in the u-v parameter space is a useful concept. The rotation number $W(V, \gamma)$ counts the number of rotations performed by $\nabla\phi(\gamma(t))$ while the point $\gamma(t)$ moves along the curve γ on the parameter space of $\vec{r}(u, v)$. In the definition of rotation number, $\nabla\phi \neq 0$ on the curve γ is assumed.

THEOREM 8.4.1. *If the rotation number of a continuous vector field $V = \nabla\phi$ along an arbitrary oriented continuous and closed curve γ, bounding a simply connected domain D, is nonzero, then there must exist at least one critical point of ϕ in domain D.*

Proof. If the continuous vector field V is never zero on D, then the rotation $W(V, \gamma)$ is zero [18]. Hence if $W(V, \gamma) \neq 0$, then V has at least one zero inside D.

$\nabla\phi$ is not continuous in a domain D if ϕ is not proper at a point in the interior of D. For intersection examples which are difficult to resolve (e.g., nearly coincident or tangent surfaces), we found that, after a small level of initial subdivision of both surfaces and elimination of nonintersecting subpatches using bounding boxes, $\nabla\phi$ is well defined and continuous at all points of the domain.

The first step in the determination of initial approximations to collinear normal points is the approximate computation of $\nabla\phi$ on a lattice of points in the parameter space of $\vec{r}(u, v)$. We use a coarse initial subdivision of the two parametric patches to provide a coarse approximation to the two surfaces. The use of bounding boxes assists in efficiently eliminating nonintersecting subpatches during this initial subdivision process. For each control point $\vec{r}_{ij}$ of the first surface, we calculate its Euclidean distance from all the control points of all the subpatches of the second surface $\vec{q}_{kl}$ and select the minimum distance pair. A pair of approximate minimum distance points is then identified in the parameter space of the two patches corresponding to the node values associated with the selected control points of the two subpatches. Our numerical experiments indicate that this procedure provides a good approximation of the nearest points after a coarse subdivision of the two surfaces [19]. As a result, $\nabla\phi$ can be approximated at a lattice of points in the u-v parameter space by computing the right-hand side of (8.2) at these points.

Once the approximate $\nabla\phi$ is determined, Theorem 8.4.1 may be used to determine initial approximations for the collinear normal points. Every four neighboring points in the lattice of points form a quadrilateral, i.e., an oriented continuous closed curve in the u-v parameter space. We found that the rotation of $\nabla\phi$ around this quadrilateral can be approximated by computing the rotation of the vectors in the corner points of the quadrilateral. If the rotation number is nonzero, then the parameter values of the center of the quadrilateral are used as initial approximations for an accurate numerical computation of a critical point. Numerical methods based on minimization and Newton iterative techniques described in [19] have been developed and successfully applied in such computation. The objective of the above method is to provide an efficient, automated way to compute collinear normal points in order to provide at least one point inside all loops and all singularities. However, there is no guarantee that, in general, at least one initial approximation exists for every collinear normal point, and that the iterative technique will converge to the appropriate collinear normal point.

To further enhance our confidence in identifying cases where the above numerical computations have missed a critical point, we may perform the following *verification* procedure based on the same concept as above but relying on accurate computation of rotation numbers as applied in appropriate subdomains of the full problem. Recall that in the simpler problem analyzed in §8.3, we subdivide (split) the domain at all available significant points obtained by some numerical computation and use a verification criterion based on derivatives of control surfaces in subdomains. This idea can be generalized to the problem addressed in this section. First, we split $\vec{r}(u,v)$ in a number of subpatches by using isoparameter lines passing through all available collinear normal points. This subdivision is performed sequentially in a manner similar to that described in §8.3.3, and the subpatches are kept in a tree structure exhibiting the same properties as those obtained in the simpler problem of §8.3. Following this subdivision, each of the resulting subpatches is examined for the possibility of a missed collinear normal point in its interior. The bounding box of each of the subpatches of the first surface is compared with the bounding boxes of the subpatches of the second surface to determine the subpatches with no intersection. This is important in reducing the number of verification tests. Tight, rectangular bounding boxes naturally oriented to the geometry of the subpatches are used to closely bound the subpatches [19], [20].

Each of the intersecting subpatches of the first surface is then examined for the existence of a missed critical point of ϕ in its interior using the rotation number of the $\nabla\phi$ on the boundary of the subpatch. If the rotation number is nonzero, then such a point may exist and needs to be identified. If the rotation number is zero, the present test is *inconclusive* since from the Poincaré index theorem [18], the rotation of $\nabla\phi$ on the boundary of a region containing one saddle and one extremum point of ϕ is zero. Numerical experiments, however, suggest that missing of two neighboring critical points of ϕ with opposite index signs is infrequent [19].

The rotation number of a vector field $V(u,v) = \{\chi, \psi\}^T$ along a closed curve γ can be determined using numerical quadrature from the Poincaré formula [18]:

$$(8.3) \qquad W(V,\gamma) = \frac{1}{2\pi} \int_\gamma \frac{\chi d\psi - \psi d\chi}{\chi^2 + \psi^2}.$$

Since the rotation along a closed curve is an integer, we only need to evaluate this integral to within an error of less than 0.5 to determine the rotation. If any point on the boundary of the subpatch (except the corners) has a null vector, the above computation cannot be performed and the test is assumed to have failed. In order to avoid integrating over critical points which are already computed and are at corners of subpatches, the integration contour may be modified using a small tolerance to avoid the critical point [19]. This modification is acceptable for intersections with isolated critical points. The tolerance indicates the separation between critical points that we want to resolve. $\nabla\phi$ at the requisite number of points on the closed boundary curve in the u-v parameter space of the first surface necessary to compute (8.3) may be evaluated by first determining the point on the second surface $\vec{q}(s,t)$ which is at a minimal distance to a given point on the boundary curve. An alternative method to compute the integrand of (8.3) is to use differential equations describing the orthogonal projection of a closed curve of $\vec{r}(u,v)$ on $\vec{q}(s,t)$ and to compute $\nabla\phi$ at the resulting points. Differential equations governing the orthogonal projection of a curve on a surface can be found in [32], [19], [33].

If the rotation number on the boundary curve is nonzero, we search locally in the subpatch for a missed critical point using the same technique explained above but applied in the smaller domain. Once a new collinear normal point is determined, the subpatch is subdivided at this point and the verification test is performed recursively in the new subpatches until it is satisfied in all of them. This process is repeated until all subpatches of the first patch satisfy this test. The collinear normal point verification method described above, although only based on a *necessary condition*, has proven successful in a large number of complex intersections by discovering collinear normal points missed in a preliminary significant point computation without substantial computational expense [19].

To provide a *sufficient condition* for the absence of loops and singularities from an intersection, convex hull properties and bounds for the partial

derivatives and normal vectors of rational B-spline surface patches may need
to be employed. The following conditions are useful in this respect. If two at
least C^1 surfaces intersect in a closed loop, then there exists a normal vector
on one surface which is *parallel* to a normal vector in the other surface [45].
Such a condition may be used to assist in selective subdivision of two surfaces
to identify all intersection segments. If bounds to the normal vector directions
of two subpatches do not intersect, these subpatches do not intersect in an
interior loop and further subdivision is unnecessary. Furthermore, if two at
least C^1 surfaces intersect in a closed loop (in both parametric spaces), then
there exists a line which is perpendicular to both surfaces (*collinear normal
vectors*), provided the inner product between any normal vector on one surface
and any other normal vector on the other surface is never zero [42]. The
absence of a collinear normal line (i.e., the absence of intersection loops) can
often be deduced even though parallel normal vectors may exist. Interval type
methods to bound a set of functions specifying collinearity to determine the
existence or not of a collinear normal have been suggested [44]. Similarly, if
two at least C^1 surfaces intersect in a closed loop, then there exists a normal
vector on one surface which is perpendicular to a tangent vector on the other
surface [43]. Cones that bound the tangent directions of all curves of constant
u or v in a rational B-spline patch (u, v bounding cones), and cones that bound
all of the normal vectors on a rational B-spline patch (normal cone) have been
proposed [43]. Using the normal cone, the tangent plane cone may be also
defined by the following property. If its vertex is translated to any point on
the surface, the tangent plane at that point will not cut the tangent plane cone.
A criterion for evaluating the existence of closed loops in the intersection curve
may be formulated as follows. If the u or v cone from one of the surfaces lies
completely within the tangent plane cone of a second surface, then all v or u
isoparameter curves of the first surface intersect the second surface at most
once. This guarantees single-valued intersection curves in v or u.

Experimentation with various possible bounds of partial derivatives and
normal vectors of rational B-spline surface patches has shown that *rectangular
pyramid bounds* may be constructed efficiently and are usually tighter than
available alternatives by a significant factor. The details of the construction
of pyramid bounds and comparisons with alternatives can be found in [19],
[20]. The center vectors of the u and v rectangular pyramids of a rational B-
spline patch may be also employed to define a coordinate system naturally
oriented to the patch. Rectangular bounding boxes of the patch position
defined in such a system are normally much smaller than bounding boxes
in an arbitrary coordinate system. Such boxes provide an efficient way to
eliminate nonintersecting subpatches generated in a subdivision process at
small computational cost. Our experimentation with intersection problems
has shown that numerical precomputation of most collinear normal points
using the rotation number criterion developed above, splitting of the patches
involved at such points, and application of sufficient conditions for loop and

singularity detection relying on rectangular pyramids offer the potential of reliable intersection component detection at small computational expense. Comparison of the relative efficiencies of algorithms to detect collinear normal points based on various combinations of necessary and sufficient conditions for the detection of such points is a subject of current investigation.

8.4.3. Tracing Schemes and Concluding Remarks. Following the computation of significant points of the intersection, two methods for tracing the intersection curve of two surfaces have been developed [19]–[21]. The first method is a subdivision and faceting technique exploiting the convex hull properties of rational B-splines in combination with a local Newton refinement of the intersection to efficiently provide high accuracy. The second method is a marching technique and is based on the solution of a set of first-order differential equations describing the zero level curves of the oriented distance function ϕ between two surfaces.

Our subdivision scheme starts after the two parametric surfaces are subdivided at the significant points and the verification criterion outlined in the previous section is possibly used. Approximate tracing of curve segments in a subdomain is based on the approximation of subpatches by their control polyhedra and the intersection of these approximations with each other. The approximation of the curve of intersection arising from the intersection of the two polyhedra is later refined using an efficient local Newton technique. The use of tight rectangular bounding boxes ensures fast elimination of nonintersecting subpatches.

Marching methods involve generation of sequences of points of an intersection curve branch by stepping from a given point on the intersection curve in a direction prescribed by the local differential geometry of the curve. Marching methods have been found to be a more efficient alternative to subdivision methods. They are simple to implement and do not generate large amounts of data during the solution process. The first difficulty with marching methods, in contrast to subdivision methods, is that they require *starting points* for every intersection branch. An important element of our work is the development of techniques to automatically determine these starting points on all connected intersection components and to identify the singular points of the curve. The second difficulty with marching methods is the selection of the *step size*. As noted in §8.2, incorrect step size may lead to erroneous connectivity of solution branches or even to endless looping in the presence of closely spaced features [16]. Most recent marching methods make use of curvature analysis or power series expansions about each point of the intersection solution to control the step size.

A new marching technique was developed which transforms the intersection curve following the (marching) problem to an equivalent, initial value ordinary differential equation problem [19], [21]. The intersection curve between two surfaces can be considered as the zero level curve of the oriented distance function between the two surfaces. In addition, if a curve $\gamma(w)$ on a surface $\vec{r}(u,v)$ and its orthogonal projection on another surface $\vec{q}(s,t)$ are identical, then curve $\gamma(w)$ is an intersection curve of the two surfaces. Orthogonal projections of curves on surfaces are studied in [32].

These two views of an intersection curve provide the tools for the development of differential equations describing the intersection curve of two surfaces. There is a large amount of literature dealing with the numerical solution of initial value problems, and there are some very good algorithms for the robust integration of the associated systems of first-order ordinary differential equations [15], [26]. These algorithms use adaptive techniques to select the proper integration order and marching step size and have proved successful in practice, particularly in areas where rapid variations and constrictions (near singularities) of the curve exist.

Marching methods, just like subdivision methods, require special attention to handle singularities reliably. In our work, we have developed techniques for marching from ordinary order 2 singular points (self-intersections) of general intersection curves $\phi = 0$ using the two asymptotic directions of the Hessian of ϕ at such points [19], [21]. The elements of the Hessian of ϕ can be computed *explicitly* in terms of partial derivatives of ϕ. The method can be extended to higher order singularities by using higher order partial derivatives of ϕ. Cusps, involving a single tangent direction, can be usually handled indirectly by arranging the sequence of marching towards, rather than away from, such points. Such an approach is expected to give correct results provided no two cusps occur within the same subdomain, or if another significant point can be detected on a branch connecting them from which marching can be initiated. Analysis of tangent directions of an intersection curve at singular points is also useful in tracing intersection curves along which the two intersecting surfaces share a common tangent plane (e.g., models involving blending surfaces). Such cases involve infinite singular points at which there is a single tangent line. When this situation arises, the convex hull/subdivision-based criteria for the detection of singular points naturally become unattractive. For such cases, we have, however, developed appropriate systems of ordinary differential equations describing the tangents to the curve, so that when a starting point is accurately known, these curves can be traced without much difficulty. An experimental implementation of these systems has led to accurate results.

The total computational cost of the intersection algorithms which use the two different tracing techniques mentioned above has been compared for a large number of complex and diverse intersections. In most instances, the marching algorithm outperforms the subdivision and Newton-based algorithm by a significant margin in accurately computing intersection curves. The following observations may also be made. The marching algorithm increases in time complexity with an increase in the complexity of the intersection curve, since it depends primarily on the number of intersection segments needed to be traced. The subdivision algorithm, however, depends less on the complexity

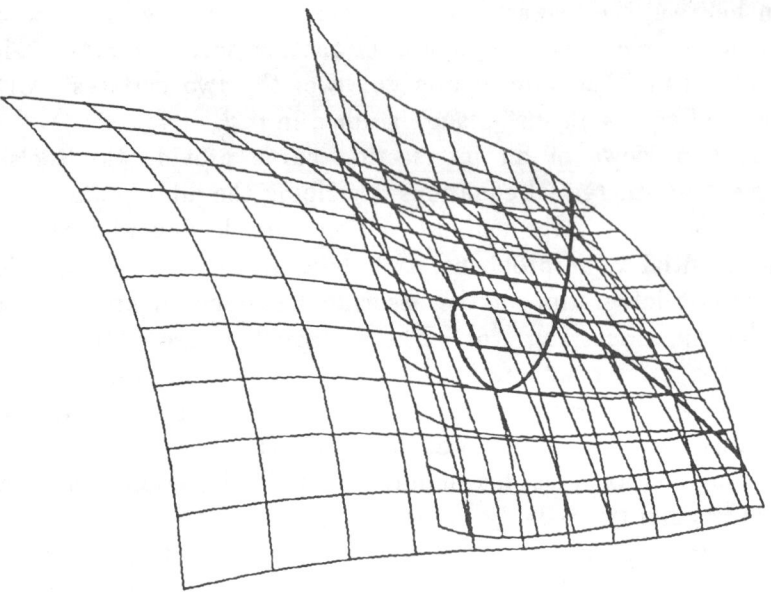

FIG. 8.4. *Bicubic-cubic/quadratic patch intersection* [19].

of the intersection and is governed primarily by the number of intersecting subpatches of the two surfaces and the number of line to plane intersections performed.

Figure 8.4 presents the intersection of a bicubic patch with a patch which is cubic in the u direction and quadratic in the v direction. The resulting intersection has two collinear normal points (one of which is also a singular point) and two border points in the parameter space of the first patch. Figure 8.5 presents the intersection of two biquartic Bézier patches (which are nearly coincident). The resulting intersection has four small loops in the parametric space of both patches.

8.5. Conclusions and Recommendations

Some important outstanding problems in the area of surface intersection computation are summarized in this section.

First, the extension of our algorithms to handle intersections of complex general parametric surfaces (such as offset, generalized cylinder, blending, and medial surfaces), surfaces arising from the solution of partial differential equations, and intersections of such surfaces with the basic algebraic and rational spline surfaces, commonly used in design, requires further study. Tracing of intersection curves by solving differential equations may be extended to handle general parametric surfaces known in a procedural fashion, intersections of surfaces exhibiting higher than tangent plane continuity, or higher order singularities. Such methods, in many respects, have been well addressed in existing state-of-the-art literature or the requisite methodologies are either of a classical

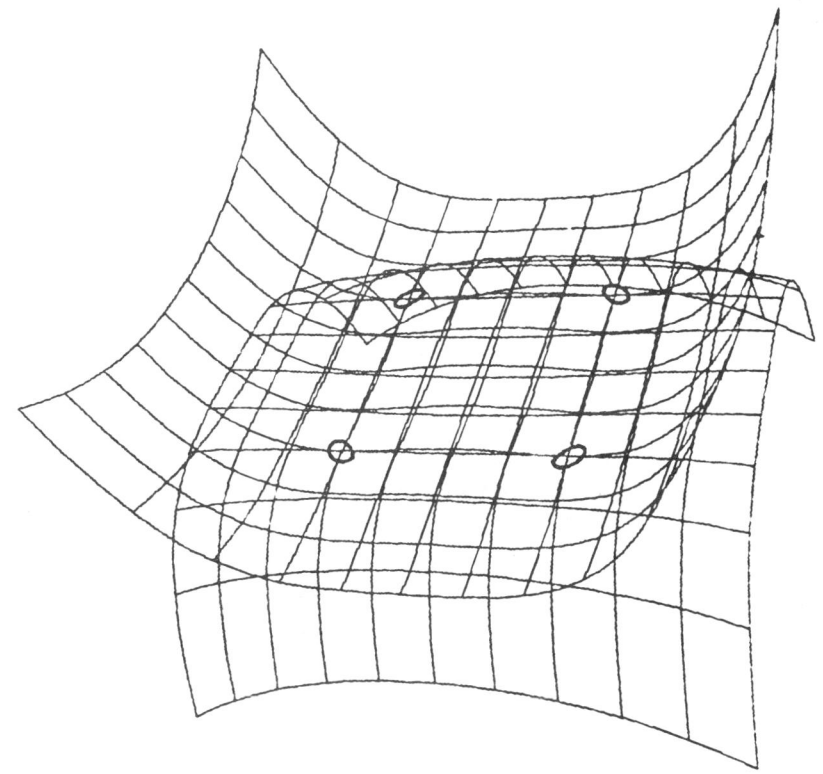

FIG. 8.5. *Intersection of two biquartic Bézier patches* [19].

nature or are currently being developed. However, a major difference between general parametric surfaces and rational B-spline surface patches, affecting the reliable interrogation of surface intersections, is the existence of a control polyhedron and associated convex hull properties and easily computable bounds for position, partial derivatives, and normal vectors in the latter class. The development of efficient nonexhaustive detection conditions for significant points and verification of computation of all connected intersection branches to such complex parametric surface intersections is the key area requiring further study. The development of tight linear or nonlinear bounds for positions, first and higher order partial derivatives, normal vectors, and other relevant properties of general parametric surfaces, which can be computed rapidly, is the dominant element in the efficient implementation of reliable feature detection conditions in such cases. Detection of surface overlap is a difficult, special instance of surface intersection which also requires further study. Distance function and vector field ideas outlined in this paper are expected to be useful in solving this complex intersection problem. The effects of floating point arithmetic on the implementation of intersection algorithms is an important area for basic research. Ways to enhance the precision of intersection computation and to monitor numerical error contamination ought to be explored. Alternate means of performing arithmetic not relying on imprecise floating point computation alone should also be explored as the demands for higher precision increase. Finally, the implementation of intersection algorithms in a parallel computation environment is an important research area, which will enhance our ability to perform design of complex systems at significantly reduced time scales in comparison to current capabilities.

The general surface to surface intersection problem, although now becoming more well understood in some of its aspects and manifestations, remains a challenging computational problem, and much research is still required to achieve a good balance between the conflicting goals of reliability, accuracy, and efficiency in algorithms attempting to solve it.

Acknowledgments

Support for the research activities of the Massachusetts Institute of Techology Ocean Engineering Design Laboratory in the surface intersection area has been provided in part by the following funding institutions: MIT Sea Grant College Program (project NA86AA-D-SG089), the Office of Naval Research (projects N00014-87-K-0462, N00014-89-J-1187 and N00014-91-J-1014), the National Science Foundation (projects DMC-8706592 and DMC-8720720), the General Electric Company, and the Doherty Professorship. The author appreciates useful discussions with Professor C. Chryssostomidis, Dr. G. A. Kriezis, Dr. P. V. Prakash, and Dr. F.-E. Wolter in the general area of shape interrogation.

References

[1] S. S. Abhyankar and C. L. Bajaj, *Automatic parametrization of rational curves and surfaces* III: *Algebraic plane curves*, Comput. Aided Geom. Des., 5 (1988), pp. 309–321.

[2] V. I. Arnold, *Ordinary Differential Equations*, MIT Press, Cambridge, MA, 1981.

[3] D. S. Arnon, *Topologically reliable display of algebraic curves*, ACM Comput. Graphics, 17 (1983), pp. 219–227.

[4] C. L. Bajaj, C. M. Hoffmann, J. E. Hopcroft, and R. E. Lynch, *Tracing surface intersections*, Comput. Aided Geom. Des., 5 (1988), pp. 285–307.

[5] R. E. Barnhill, G. Farin, M. Jordan, and B. R. Piper, *Surface/surface intersection*, Comput. Aided Geom. Des., 4 (1987), pp. 3–16.

[6] R. E. Barnhill and S. N. Kersey, *A marching method for parametric surface / surface intersection*, Comput. Aided Geom. Des., 7 (1989), pp. 257–280.

[7] H. Blum, *Biological shape and visual science (part* I), J. Theoret. Biol., 38 (1973), pp. 205–287.

[8] Y. J. Chen and B. Ravani, *Offset surface generation and contouring in computer-aided design*, J. Mech. Trans. Automat. Des., ASME Trans., 109 (1987), pp. 133–142.

[9] K. P. Cheng, *Using plane vector fields to obtain all the intersection curves of two general surfaces*, in Proceedings of the Theory and Practice of Geometric Modeling, W. Strasser and H. Seidel, eds., Springer-Verlag, New York, 1988, pp. 187–204.

[10] C. Chryssostomidis and N. M. Patrikalakis, *Geometric modeling issues in computer aided design of marine structures*, Marine Tech. Soc. J., 22 (1988), pp. 15–33.

[11] Y. de Montaudouin, W. Tiller, and H. Vold, *Applications of power series in computational geometry*, Comput. Aided Des., 18 (1986), pp. 514–524.

[12] M. P. do Carmo, *Differential Geometry of Curves and Surfaces*, Prentice-Hall, Englewood Cliffs, NJ, 1976.

[13] R. T. Farouki, *The characterization of parametric surface sections*, Computer Vision, Graphics, and Image Processing, 33 (1986), pp. 209–236.

[14] R. T. Farouki and V. T. Rajan, *On the numerical condition of polynomials in bernstein form*, Comput. Aided Geom. Des., 4 (1987), pp. 191–216.

[15] C. W. Gear, *Numerical Initial Value Problems in Ordinary Differential Equations*, Prentice-Hall, Englewood Cliffs, NJ, 1971.

[16] A. Geisow, *Surface interrogations*, Ph.D. Thesis, School of Computing Studies and Accountancy, University of East Anglia, Norwich, England, July 1983.

[17] C. M. Hoffmann, *Geometric and Solid Modeling: An Introduction*, Morgan Kaufmann Publishers, San Mateo, CA, 1989.

[18] M. A. Krasnoselskiy, A. I. Perov, A. I. Povolotskiy, and P. P. Zabreiko, *Plane Vector Fields*, Academic Press, New York, 1966.

[19] G. A. Kriezis, *Algorithms for rational spline surface intersections*, Ph.D. Thesis, Dept. of Ocean Engineering, Massachusetts Institute of Technology, Cambridge, MA, March 1990.

[20] G. A. Kriezis and N. M. Patrikalakis, *Rational polynomial surface intersections*, in Proceedings of the 17th ASME Design Automation Conference, Miami, FL, G. Gabriele, ed., ASME, New York, September 1991.

[21] G. A. Kriezis, N. M. Patrikalakis, and F. E. Wolter, *Topological and differential equation methods for surface intersections*, Comput. Aided Des., (1991), to appear.

[22] G. A. Kriezis, P. V. Prakash, and N. M. Patrikalakis, *A method for intersecting algebraic surfaces with rational polynomial patches*, Comput. Aided Des., 22 (1990), pp. 645–654.

[23] J. M. Lane and R. F. Riesenfeld, *A theoretical development for the computer display and generation of piecewise polynomial surfaces*, IEEE Trans. Pattern Analysis and Machine Intelligence, 2 (1980), pp. 35–46.

[24] J. Z. Levin, *Mathematical models for determining the intersections of quadric surfaces*, Computer Vision, Graphics, and Image Processing, 11 (1979), pp. 73–87.

[25] R. P. Markot and R. L. Magedson, *Solutions of tangential surface and curve intersections*, Comput. Aided Des., 21 (1989), pp. 421–429.

[26] NUMERICAL ALGORITHMS GROUP, *NAG Fortran Library Manual*, Mark Thirteenth Edition, Oxford, England, 1989.

[27] N. M. Patrikalakis, *Shape interrogation*, in Proceedings of the 16th Annual MIT Sea Grant College Program Lecture and Seminar, Automation in the Design and Manufacture of Large Marine Systems, Cambridge, MA, October 1988, C. Chryssostomidis, ed., Hemisphere Publishing, New York, 1990, pp. 83–104.

[28] N. M. Patrikalakis and H. N. Gursoy, *Shape interrogation by medial axis transform*, in Proceedings of the 16th ASME Design Automation Conference: Advances in Design Automation, Computer Aided and Computational Design, Chicago, IL, Vol. I, B. Ravani, ed., ASME, New York, September 1990, pp. 77–88.

[29] N. M. Patrikalakis and G. A. Kriezis, *Representation of piecewise continuous algebraic surfaces in terms of B-splines*, The Visual Computer, 5 (1989), pp. 360–374.

[30] N. M. Patrikalakis and P. V. Prakash, *Computation of algebraic and polynomial parametric surface intersections*, Technical Report MITSG 87-19, MIT Sea Grant College Program, Cambridge, MA, 1987.

[31] ——, *Surface intersections for geometric modeling*, J. Mech. Des., ASME Trans., 112 (1990), pp. 100–107.

[32] J. Pegna, *Interactive design of curvature continuous fairing surfaces*, in Proceedings of the 8th International Conference on OMAE, the Hague, the Netherlands, N. M. Patrikalakis, ed., Vol. VI, ASME, New York, March 1989, pp. 191–198.

[33] J. Pegna and F. E. Wolter, *Designing and mapping trimming curves on surfaces using orthogonal projection*, in Proceedings of the 16th ASME Design Automation Conference: Advances in Design Automation, Computer Aided and Computational Design, Chicago, IL, Vol. I, B. Ravani, ed., ASME, New York, September 1990, pp. 235–245.

[34] C. S. Petersen, *Adaptive contouring of three-dimensional surfaces*, Comput. Aided Geom. Des., 1 (1984), pp. 61–74.

[35] P. V. Prakash, *Computation of surface-surface intersections for geometric modeling*, Ph.D. Thesis, Dept. of Ocean Engineering, Massachusetts Institute of Technology, Cambridge, MA, May 1988.

[36] P. V. Prakash and N. M. Patrikalakis, *Surface-to-surface intersections for geometric modeling*, Technical Report MITSG 88-8, MIT Sea Grant College Program, Cambridge, MA, 1988.

[37] J. R. Rossignac and A. A. G. Requicha, *Piecewise-circular curves for geometric modeling*, IBM J. Res. Develop., 31 (1987), pp. 296–313.

[38] R. F. Sarraga, *Algebraic methods for intersections of quadric surfaces in GMSOLID*, Computer Vision, Graphics, and Image Processing, 22 (1983), pp. 222–238.

[39] T. W. Sederberg, *Planar piecewise algebraic curves*, Comput. Aided Geom. Des., 1 (1984), pp. 241–255.

[40] ——, *Piecewise algebraic surface patches*, Comput. Aided Geom. Des., 2 (1985), pp. 53–59.

[41] T. W. Sederberg, D. C. Anderson, and R. N. Goldman, *Implicit representation of parametric curves and surfaces*, Computer Vision, Graphics, and Image Processing, 28 (1984), pp. 72–84.

[42] T. W. Sederberg, H. N. Christiansen, and S. Katz, *An improved test for closed loops in surface intersections*, Comput. Aided Des., 21 (1989), pp. 505–508.

[43] T. W. Sederberg and R. J. Meyers, *Loop detection in surface patch intersections*, Comput. Aided Geom. Des., 5 (1988), pp. 161–171.

[44] T. W. Sederberg and T. Nishita, *Direct approximation of surface patch intersection curves*, Tecnical Report, Brigham Young University, Provo, UT, January 1989.

[45] P. Sinha, E. Klassen, and K. K. Wang, *Exploiting topological and geometric properties for selective subdivision*, in Proceedings of Association for Computing Machinery Symposium on Computational Geometry, ACM, New York, 1985, pp. 39–45.

[46] B. J. Solomon, *Surface intersection for solid modelling*, Ph.D. Thesis, Clare College, University of Cambridge, Cambridge, England, 1985.

[47] S. W. Thomas, *Modeling volumes bounded by B-spline surfaces*, Ph.D. Thesis, Department of Computer Science, University of Utah, Salt Lake City, UT, June 1984.

[48] T. Varady, *Surface-surface intersections for double-quadratic parametric patches in a solid modeller*, in Proceedings of the U.K.–Hungarian Seminar on Computational Geometry for CAD/CAM, Cambridge University, Cambridge, England, 1983.

[49] F. E. Wolter, *Cut loci in bordered and unbordered Riemannian manifolds*, Ph.D. Thesis, Department of Mathematics, Technical University of Berlin, Berlin, Germany, December 1985.

Parametric Surface Intersections

K. Y. Wang

9.1. Introduction

Intersections between surfaces have been a subject of great interest [9]. The success of many industrial applications, such as tool path generation for numerical controls, surface blending, interference detection, and surface trimming, often depends on the generation of intersection curves between sculptured surfaces. Depending on the purpose of the applications, either parametric or algebraic expressions can be used to describe a surface. In this article, we will concentrate on the intersection techniques for the parametric surfaces.

Analytical approaches have been the traditional way to solve the intersection problems. It has the potential payoff of being able to represent the intersection curve exactly and thus possibly reduce the size of data, as compared to the approximation methods. Implicitizing both surfaces and solving the resulting simultaneous algebraic equations by elimination is one possible approach [7], [10]. The number of variables can be reduced, although the advantage may be offset by raising the degrees of the algebraic polynomials to be solved. Studies have also been carried out on a possible method to replace the high degree intersection curves with lower degree, locally implicit approximations [2].

An alternative method is to implicitize only one of the surfaces and transform the problem to the intersection between a Bézier surface and a plane [4], [11]. The intersection curve can be reduced to the description of a planar algebraic curve within a rectangular domain described by just two parameters. Nonetheless, the algebraic expressions are again polynomials of very high degrees. Solving such high degree polynomials again becomes a prohibitive task. Therefore, even though the analytical approach may provide the user with the precise nature of the intersection curves and is considered a powerful tool in handling special geometries such as conic surfaces, the difficulty associated with solving high degree polynomials makes it unsuitable to be extended to handle the more general sculptured surfaces. As a result, approximations become an alternative that at least provides some workable solutions [2], [5], [6], even though the results may be incomplete.

Another widely used approach is the facet method, which simplifies the geometry by linear segments [3], [5], [6]. A marching [1] or subdividing [11] algorithm usually follows to enhance the efficiency of the algorithm. The biggest drawback of the approach is the danger of topological inconsistency due to improper tessellation [13], [14], which is hard to detect and correct even when enhanced by other methods. It is not uncommon that facet approximation fails in the near tangent situations, as illustrated in Fig. 9.1.

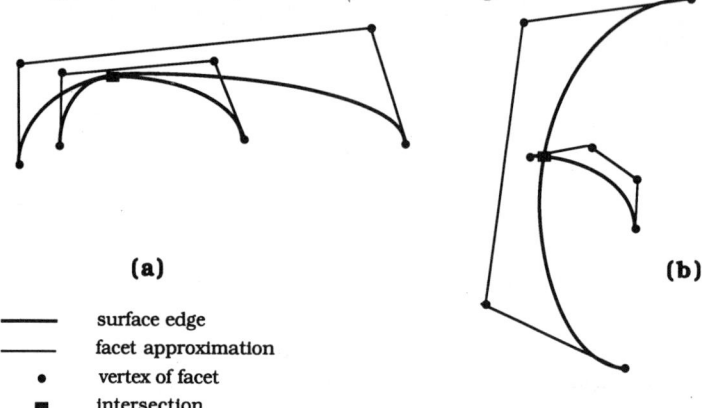

(a)   **(b)**

————   surface edge
————   facet approximation
  •   vertex of facet
  ■   intersection

FIG. 9.1. *Cases where intersection may be missed by facet approximations.* (a) *Surfaces nearly tangent at intersection;* (b) *edge of one surface is close to intersection.*

The goal of this work is to try to address some of the difficult issues mentioned above, and through approximations provide a compromise between accuracy and efficiency. Intersections between a curve and a surface are developed as the foundation of the solution. Curves are fitted through exact intersection points as a representation for the intersection. The adaptive control algorithm, which continuously evaluates the accuracy of the intersection curve and generates more sampling points only on demand, is a feature that helps in reducing the data size. The distance evaluation function will be introduced for the purpose of identifying intersections, and seems to be very useful for the less obvious cases such as tangential contact and small loop intersections. A few examples will be given at the end to illustrate the principles.

9.2. Approach

The method presented here can be divided into the following steps:

1. searching for sampling points on the intersection curves,

2. establishing the relations among the sampling points through sorting,

3. curve fitting, and

4. verification and adaptive correction.

The first step requires algorithms that accurately compute sampling intersection points between the two surfaces. The second step is sorting, which organizes the sampling points into different groups so that curves can be fitted through them. Both of these steps are exact representations of the intersection. The fidelity of the fitted curve in step 3 depends on the number of points available for fitting as well as the way the curve is created. A verification procedure is implemented to correct possible errors incurred from steps 1–3, and to provide a mechanism to automatically adjust interior points for better accuracy.

9.2.1. Arc/Patch Intersection. The process begins by checking if the two surfaces overlap. Bounding volumes are constructed and compared to eliminate unnecessary computations [6]. One example is applying the Bézier convex hull property for the rational B-spline surfaces. The search continues if the bounding boxes built around the surface patches intersect each other. The sampling intersection points are defined as the intersections between one patch and selected parametric curves (isoparametrics) of the other patch. The patch border is always included in the sampling, and its intersections with the other patch are called the border points. The number of isoparametrics to be used usually affects the accuracy of the solution. More isoparametrics mean smaller spacing between them and will probably generate more intersection points available for fitting, although at the expense of more computing time. On the other hand, the employment of adaptive control, which generates more isoparametrics on demand to enhance accuracy, and the distance function, which enables the system to identify tiny closed loops or singularities (both to be discussed later), have lessened the need to use a large number of isoparametrics. We have found that it is usually sufficient to choose initially between two and five isoparametrics in each parametric coordinate.

A list of trial points from the patch and the isoparametric is used as initial points for computing the intersection between the arc and the patch. The trial points are usually distributed uniformly over the parametric spaces. Increasing the number of trial points will normally reduce the possibility of an incorrect count, but also slow down the computation. Because of the importance of border points in determining the sorting pattern, it is recommended that more trial points be used when computing the intersections between the border isoparametric and the patch. An incorrect count in the middle of the surface is usually less critical since it can be caught and corrected during verification. The exception is missing a critical point where two surfaces touch at a single point. However, the distance check should identify such points and reduce the ambiguity. The points are ranked according to the distance between them and then selected as the initial points.

A rational surface r of degree m and n in parametric coordinates u and v, respectively, can be expressed as

(9.1) $$r(u,v) = \frac{\sum_{i=0}^{m}\sum_{j=0}^{n} a_{ij}u^i v^j}{\sum_{i=0}^{m}\sum_{j=0}^{n} b_{ij}u^i v^j},$$

where a_{ij} and b_{ij} are the coefficients. As illustrated in Fig. 9.2, $p(u)$ and $q(s,t)$ are the initial points from the arc and the patch, respectively. Point r is at the intersection between a line tangent to the isoparametric at p and a plane tangent to the patch at q. A set of new parametric values $\tilde{u}$, $\tilde{s}$, and $\tilde{t}$ for r is calculated from

(9.2) $$\tilde{u} = u + du, \quad \tilde{s} = s + ds, \quad \tilde{t} = t + dt,$$

where the incremental parametric values du, ds, and dt are obtained by solving the following vectorial equations:

(9.3) $$r = p(u) + \frac{dp(u)}{du}du,$$

(9.4) $$r = q(s,t) + \frac{\partial q(s,t)}{\partial s}ds + \frac{\partial q(s,t)}{\partial t}dt.$$

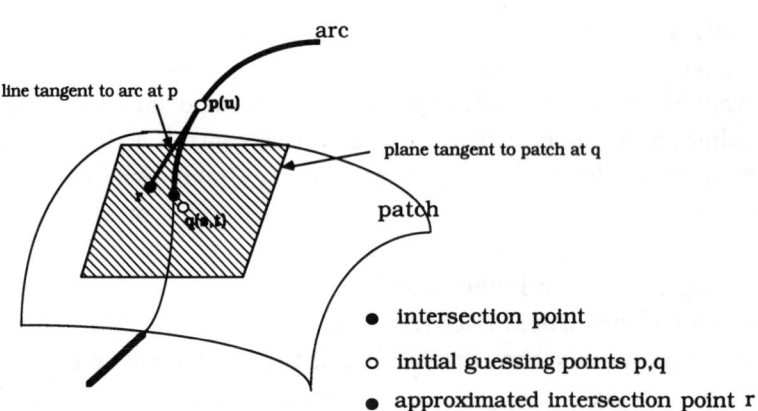

arc

line tangent to arc at p

$p(u)$

plane tangent to patch at q

$q(s,t)$

patch

● intersection point

○ initial guessing points p,q

● approximated intersection point r

FIG. 9.2. *Curve-surface intersection.*

Points that correspond to these parametric values, i.e., $p(\tilde{u})$ and $q(\tilde{s},\tilde{t})$, then become the starting points for the next iteration. The iteration terminates if $(p-q)$ is smaller than the specified tolerance. Typically, an accuracy of 10^{-10} can be achieved within a few iterative steps. The search is repeated for other candidates from the sampling list until no new intersection points can be found. Although it is difficult to be absolutely certain that all intersection points have been caught, the subsequent computations in sorting and verification provide the possible remedy by checking the number of border points and examining the accuracy. More sampling points will be generated adaptively and the computation will be repeated if such tests fail.

The approach is applicable whenever there are a finite number of intersection points between the curve and the surface, but might become inefficient or fail for cases having coincidence, where there are an infinite number of intersection points between the two geometries. Such conditions are identified beforehand by checking the distance from the end points of the curve to the surface, and the distance from the curve to the surface borders. More information, such as curve tangent direction and surface normals, will be generated and compared if the distance turns out to be less than the tolerance. The intersection curve in this case is the range of the coincidence and can be calculated based on this distance evaluation.

If the points $p(u, v)$ and $q(s, t)$ are considered identical in space, i.e., the distance between them is smaller than the tolerance, the tangent direction t of the intersection curve can be expressed as

$$(9.5) \qquad t = (p_u \times p_v) \times (q_s \times q_t),$$

which is the cross product of the normal vectors from each surface. If two surfaces are tangent to each other, then both normal vectors at the contact point will project to the same direction, and the cross product is the zero vector. This is the case for singularities (single point contact or intersecting and being tangent simultaneously) and requires special treatment in sorting. It should be pointed out that care must be exercised in computing and comparing the surface normals because quite a few mathematical operations are involved and the numerical round-off errors could be accumulated quickly. Using double precision computation is the minimal requirement for such operations.

There are situations when the surface normals are difficult to define, such as at the apex of a cone. This problem is circumvented by checking the neighboring intersection points. A singular point is identified when the intersection point is an isolated one (such as the cone touches a surface at the apex). Otherwise, the normal direction at the point can be defined asymptotically from its neighborhood (such as the cone intersects with a plane perpendicular to its base and passes its apex).

9.2.2. Sorting. The computed sampling points are scattered over the parametric domain and need to be sorted into ordered paths for curves to follow. Without any singularity or closed loop, the number of border points should be twice the number of intersection curves in the patch, with all of the interior points belonging to any one of the curves. Singularities on the borders can be identified by examining the surface normals, and the existence of a closed loop can be detected beforehand by the distance check explained in §9.2.4. The sorting for the intersection points, expressed as $p(u, v)$ or $q(s, t)$, is performed on the two-dimensional (2D) parametric spaces (u, v) or (s, t). The tangent vector t should also be mapped back to the parametric coordinates. When expressed as an angle θ from the u axis, it is

$$(9.6) \qquad \theta = \arctan \frac{(t \cdot p_v)(p_u \cdot p_u) - (t \cdot p_u)(p_u \cdot p_v)}{(t \cdot p_u)(p_v \cdot p_v) - (t \cdot p_v)(p_u \cdot p_v)}.$$

Similar expressions can be derived for the mapping of point $q(s, t)$ to (s, t).

The system automatically generates more points for sorting whenever uncertainty arises. Ambiguity in sorting direction, a noneven number of border points, and more than two border points existing could be the sources of uncertainty. An example is shown in Fig. 9.3 where three possibilities exist: a–b–c–d and e–f–g–h, or a–b–f–e and d–c–g–h, or a–b–g–h and e–f–c–d. Here a, d, e, and h are known to be the border points from the previous calculations. Intersection points are added adaptively within the zone of confusion, b–c–g–f, and their tangent directions will tell which case should apply. In fact, the directions at the border points already give us a clue whether it is case 1 or case 2. For the third case, the curves cross each other, indicating the existence of a singularity. This condition can be identified by examining the surface normals, as they should be collinear at a singular point s. The singularity will have to be found by the adaptive search within the range b–c–g–f. The range is squeezed successively until the surface normal check is satisfied. Except for the case of coincidence, a marching algorithm [11] can also be employed as an alternative. Points not sorted to the right branches can be detected during verification because the fitted curves will be found to be too far from both surfaces.

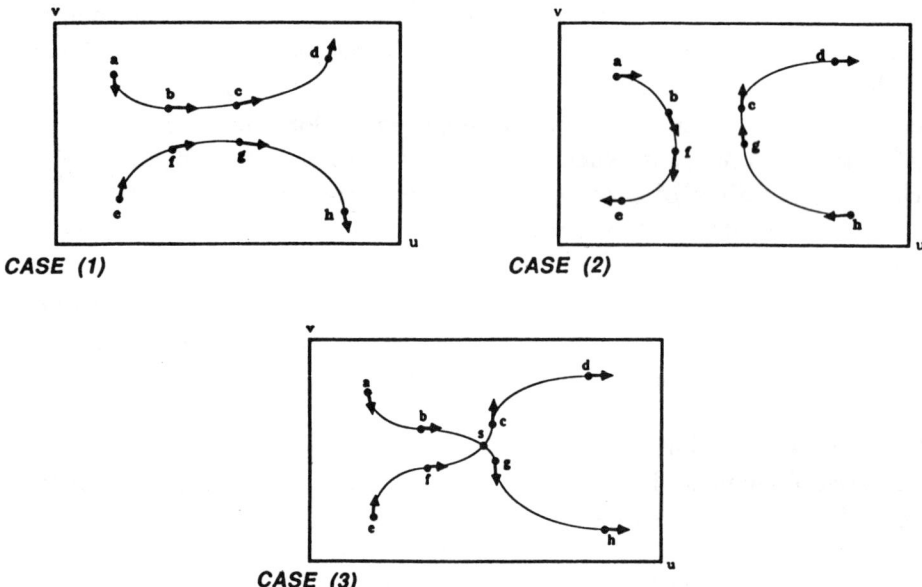

FIG. 9.3. *Tangent directions at the intersection points are useful in sorting.*

9.2.3. Curve Fitting and Verification. A theoretically exact intersection curve may require a very high degree polynomial for representation [10], which makes it impractical to implement. Therefore, curve fitting with lower degree polynomials is an acceptable compromise. An acceptable approximation requires that the maximum distance from the curve to the surfaces always be within the tolerance, and the tangent directions be preserved. Techniques of fitting approximation can be found in the work of Waggenspack [14]. An adaptive control is implemented so that more intersection points are added whenever the distance is larger than the tolerance. Errors incurred previously in the arc/patch intersection or sorting will show up during the verification and prompt repeated calculations and corrections.

9.2.4. Assurance of Intersection. A concern that may be raised is the selection of the sampling isoparametrics. It is possible that none of the selected isoparametrics from one patch intersect the other patch, and as a result the intersection is missed. One possible situation is when two surfaces intersect in a loop so small that none of the isoparametrics cross it. To overcome this difficulty, a local minimum/maximum distance check can be performed beforehand to eliminate the uncertainty. The local minimum/maximum distance occurs at locations where both surfaces are parallel to each other. There are three possibilities when this occurs: surfaces that are separated (minimum distance exists), intersected (maximum distance exists), or tangent to each other (zero distance between them). As indicated in Fig. 9.4, these cases can easily be identified by examining the direction of a line pq joining surfaces A and B at locations where the minimum or maximum distance occurs. Intersection is guaranteed when pq points to a direction opposite the surface normal of A, as indicated in case (a). Separation is just the opposite, as indicated in case (b). The special case of tangential contact, as shown in case (c), occurs when the minimum distance is zero and surface normals are parallel. The method presented in [8], under the terminology of vector field, actually addressed the issue of tangential contacts. Such points are identified as having zero distances between them and having collinear surface normals. The zeros of the vector field then become the tangential contact points.

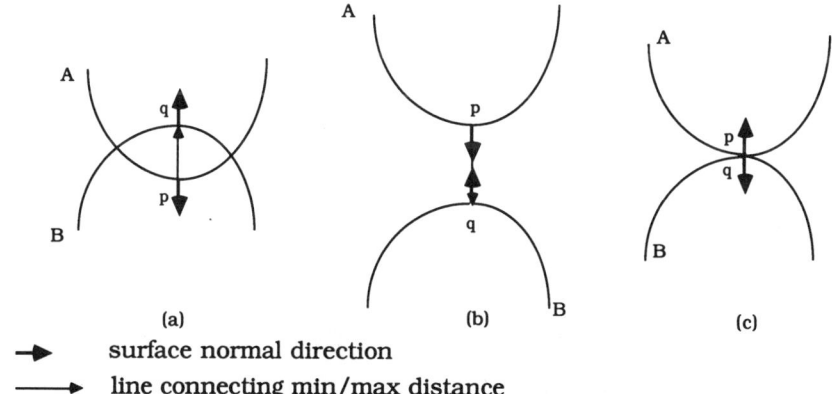

surface normal direction
line connecting min/max distance

FIG. 9.4. *Distance vector between two surfaces is an indicator for their relative positions:* (a) *intersection;* (b) *separation;* (c) *tangent.*

Another way of looking at case (a) is that there exists a zero distance, but the surface normals are not collinear. As the distance evaluation is not restricted to the selection of isoparametrics, the existence of intersections can be detected. Thus, if the minimum distance is larger than the prescribed tolerance, there will be no intersections. On the contrary, intersection exists when the minimum distance is smaller than the tolerance. As shown in the next section, the solutions of the distance equations are selective points on the loop. With such points known, it is not difficult to choose other surrounding isoparametrics and generate more sampling points for curve fitting.

9.2.5. Minimum Distances. Let A and B be surfaces defined by parameters (u, v) and (s, t), respectively. The distance D between any point $p(u, v)$ on A and $q(s, t)$ on B can be expressed as

$$(9.7) \qquad D = [(p(u, v) - q(s, t)) \cdot (p(u, v) - q(s, t))]^{1/2}.$$

If p, q are the pair of points where the shortest distance between A and B occurs, the following conditions should be satisfied:

$$(9.8) \qquad F \equiv \frac{\partial D}{\partial u} = 2p_u \cdot (p - q) = 0,$$

$$(9.9) \qquad G \equiv \frac{\partial D}{\partial v} = 2p_v \cdot (p - q) = 0,$$

$$(9.10) \qquad H \equiv \frac{\partial D}{\partial s} = -2q_s \cdot (p - q) = 0,$$

$$(9.11) \qquad I \equiv \frac{\partial D}{\partial t} = -2q_t \cdot (p - q) = 0.$$

Let points $p(u_0, v_0)$ and $q(s_0, t_0)$ be in the vicinity of p and q, say from previous iteration. Taking a linear approximation, the above equations can be rearranged as

$$(9.12) \qquad F = F_0 + dF(du, dv, ds, dt) = 0,$$

$$(9.13) \qquad G = G_0 + dG(du, dv, ds, dt) = 0,$$

$$(9.14) \qquad H = H_0 + dH(du, dv, ds, dt) = 0,$$

$$(9.15) \qquad I = I_0 + dI(du, dv, ds, dt) = 0.$$

The coefficients F_0, G_0, H_0, I_0 represent surface properties evaluated at (u_0, v_0) and (s_0, t_0). The four unknowns du, dv, ds, and dt can be obtained by solving the above four simultaneous linear equations. The new parametric values for the next iteration are computed from

$$(9.16) \qquad u = u_0 + du, \quad v = v_0 + dv, \quad s = v_0 + ds, \quad t = v_0 + dt.$$

The iteration is repeated until du, dv, ds, and dt are smaller than the tolerance described earlier. It is obvious that the solution corresponds to two situations:

either the derivatives of p and q are perpendicular to the vector $(p - q)$, or $p = q$. The former case applies when the surface normals are parallel (the two geometries do not intersect there), representing a locally minimum or maximum distance case. The latter situation indicates the existence of an intersection between the two surfaces. A check of the surface normals at the point should be sufficient to tell whether the two surfaces touch tangentially at that single point (a singularity), or the intersection is a curve.

9.3. Examples and Discussion

Four examples are provided in this section to illustrate the algorithms described above. The first example shows the intersection between a cylinder and a cone. The two geometries are plotted in Fig. 9.5(a). A parametric line on the conic surface, joining the apex and the base, is partially coincident with the side of the cylinder. The coincidence is calculated according to the method given earlier in §9.2.1. Depending on how the cylinder is constructed, the sorting pattern could be quite different. Two possible situations are given in Fig. 9.5(b) and Fig. 9.5(c) which show the close-up view of the two patches consisting of the coincidence segment a–b. In Fig. 9.5(b), the coincidence occurs at the border of the conical patch, and the other branch of the intersection curve c–d ends in the middle of a–b. In Fig. 9.5(c), the coincidence is not part of the border, and the intersection curves cross each other at the singular point e. Since segment a–b is obtained from coincidence calculation, no sorting is needed there. The searching for d in Fig. 9.5(b), which is a border point as well as a singular point, is prompted by the condition that only one nonsingular border point c can be found in the calculations, and that the branch has to march to somewhere on the border along its tangent direction. More points are generated adaptively along that direction so that d can be approached asymptotically. The situation in Fig. 9.5(c) is the same as case 3 of Fig. 9.3. The singular point e is found by adaptively adding more isoparametrics between a–b and c–d. The complete intersection curves are shown in Fig. 9.5(d).

In Fig. 9.6, the intersection between two sculptured surfaces, each consisting of over 1000 patches, is computed. The problem is complicated by the large number of surface patches to be dealt with as well as by the possibility of multiple curve/surface intersections and discontinuities in the slope of the intersection curves. Figure 9.6(a) shows the two airfoil-shaped geometries, and Fig. 9.6(b) shows the two intersection curves. Each curve is a closed loop. The short, straight segment of the curves is the intersection between the flat end-cap sections. The accuracy is illustrated in Fig. 9.6(c) where the abscissa represents the global parametric value of the intersection curve, and the ordinate is the distance from the point on the curve to the surface. In this example, only patch borders are used as isoparametrics, yet the accuracy is better than 10^{-4}. When the adaptive feature is applied, more sampling points are generated at the locations where the distance is larger than the tolerance, and a more accurate intersection curve can be created.

(a)

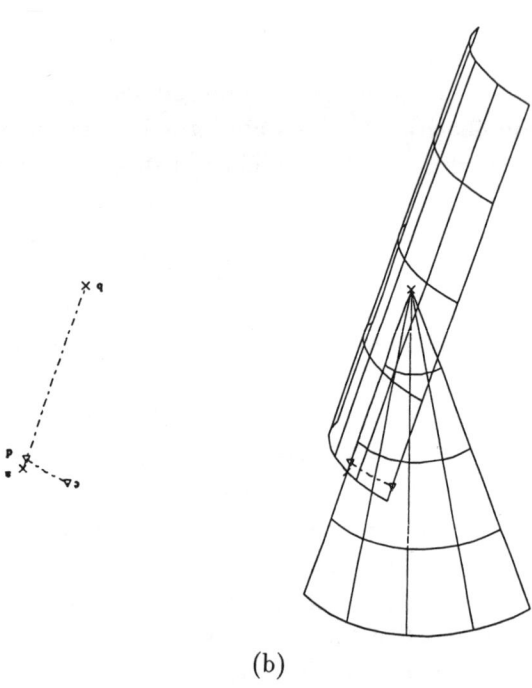

(b)

FIG. 9.5. *Intersections between a cylinder and a cone:* (a) *the two geometries are partially coincident on one edge;* (b) *close-up view of the patches containing the coincidence;* (c) *same as* (b), *but the cylinder is slightly rotated;* (d) *the intersection curve.*

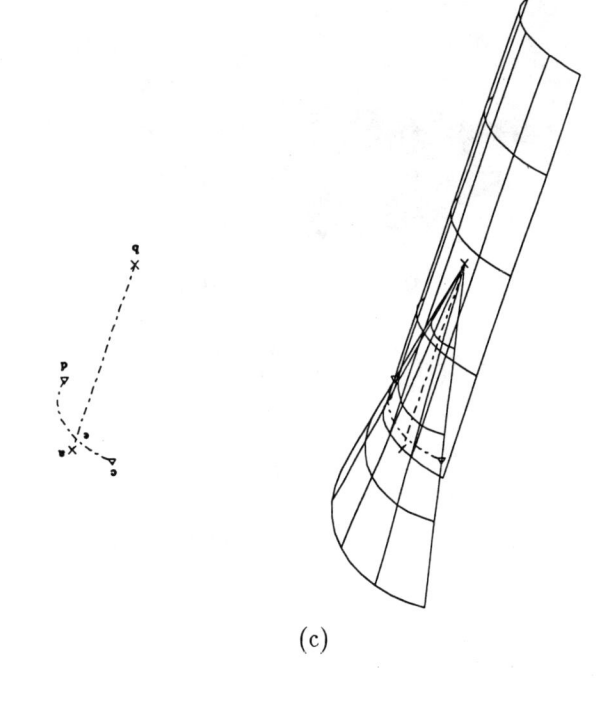

(c)

(d)

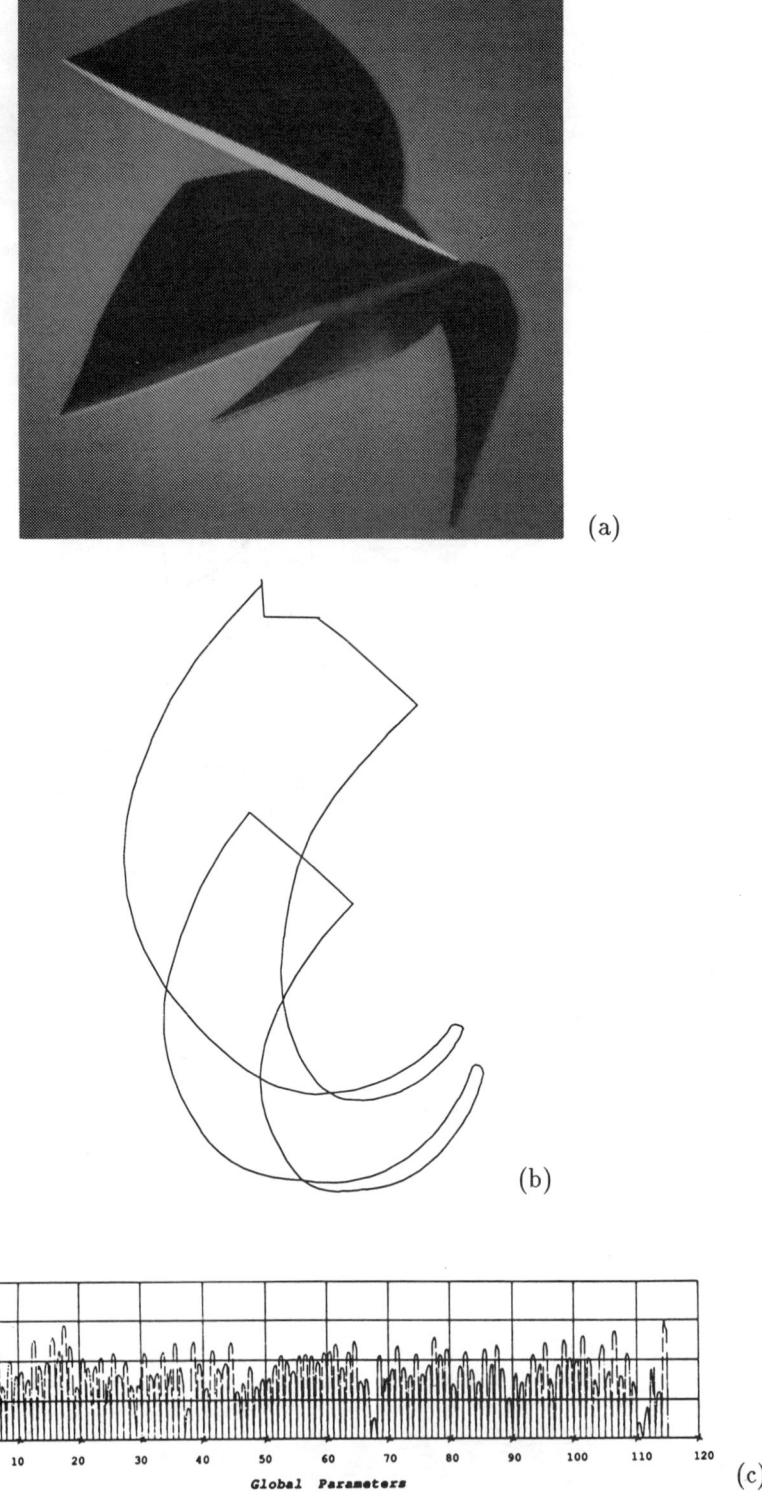

FIG. 9.6. *Intersections between two primitives:* (a) *the geometries, each consists of over* 1000 *patches;* (b) *the two intersection loops;* (c) *verification of the accuracy.*

Singularity exists in the intersection curves shown in Fig. 9.7. The singular point is where the two surfaces are tangential to each other, and the surface normal vectors are collinear. A sphere and a cylinder with a smaller diameter are used as examples. As revealed in Fig. 9.7(a), the sphere cuts through the cylinder and is tangent to it on one side. It is well known theoretically that the intersection curve is a self-intersecting loop, as shown in Fig. 9.7(b), viewed from a direction normal to the cylinder axis. The singular point on the intersection curve, shown as a diamond symbol on both figures, is where the two geometries are tangent to each other. The cylinder has been deliberately rotated by a small angle about its axis so that the singularity does not lie on the patch borders. This will make the task much more difficult and provides an opportunity to test the reliability of the sorting procedure. The sampling intersection points for the spherical patch containing this singular point are mapped back to the parametric coordinates, and are plotted on Fig. 9.7(c). Without knowing the exact location of the singularities, it will be difficult to group these points into proper branches. Inadequate information may lead to the conclusion that the intersections are two separate arc segments, as exemplified in Fig. 9.7(d). The verification will indicate that such sortings are incorrect because the distance near the singular point to both surfaces becomes too big. The exact location of the singularity is calculated by adaptively narrowing down the searching domain until collinear surface normals at the intersection point can be reached.

The next example deals with the case where the intersection loop is tiny compared to the dimensions of both surfaces. For the purpose of easy illustration, a cube and a sphere are used. As shown in Fig. 9.8(a), the sphere is centered at (0.01,0.02,0.01) with unity as the radius, and the cube is centered at (2.0095,0,0) with 2 as the length for every side. The sphere is also rotated slightly about its center so that the intersection loop is completely inside one of the patches. Such manipulation would make the computation more difficult because none of the patch borders cross the other surface and there is no clear indication of intersection, unless the minimum distance check is performed. The intersection loop, which is a circle, is shown in Fig. 9.8(b). The solid boundaries represent the edges of the cube on which the intersection curve lies. The spacing between the dashed grid is 0.1. There is no guarantee that the intersection can be found by tracing arbitrarily selected isoparametrics, unless very small sampling intervals are used. A distance check, as outlined earlier, indicates the existence of the intersection and provides starting points for searching. Other surrounding isoparametrics are thus generated adaptively to complete the loop.

9.4. Concluding Remarks

The method described in this article is aimed at solving the problem of intersection between sculptured parametric surfaces. The intersections are established by fitting curves through exact points which are the intersections

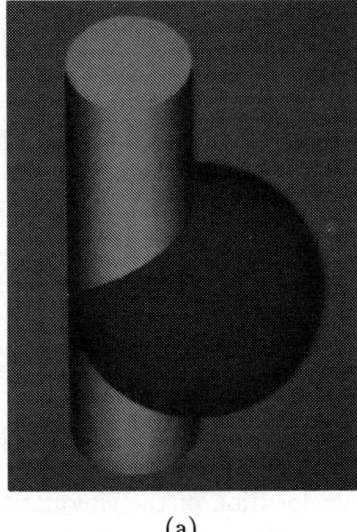

(a)

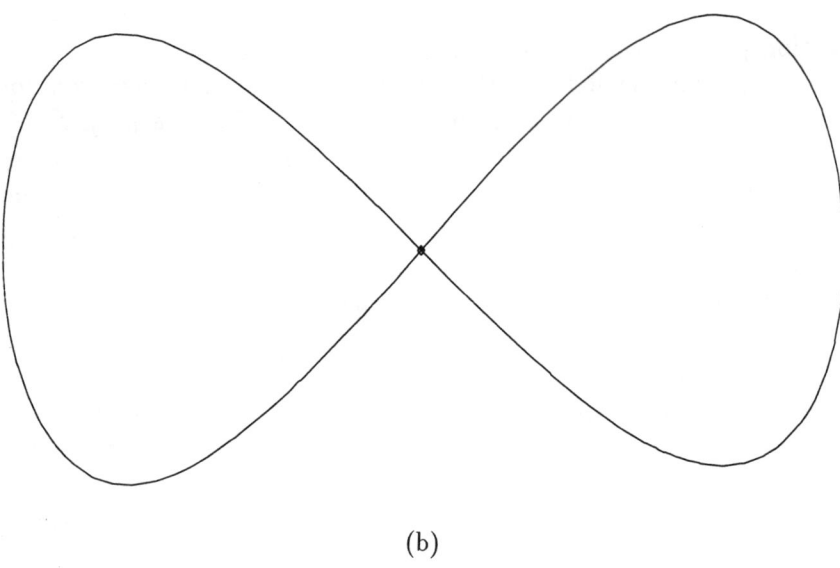

(b)

FIG. 9.7. *Intersections of a sphere and a cylinder: (a) the two geometries; (b) the intersection curve, the diamond symbol is the singular point; (c) intersection points on the parametric space of the spherical patch; (d) the error may arise if the singular point is missed, shown here as dashed lines.*

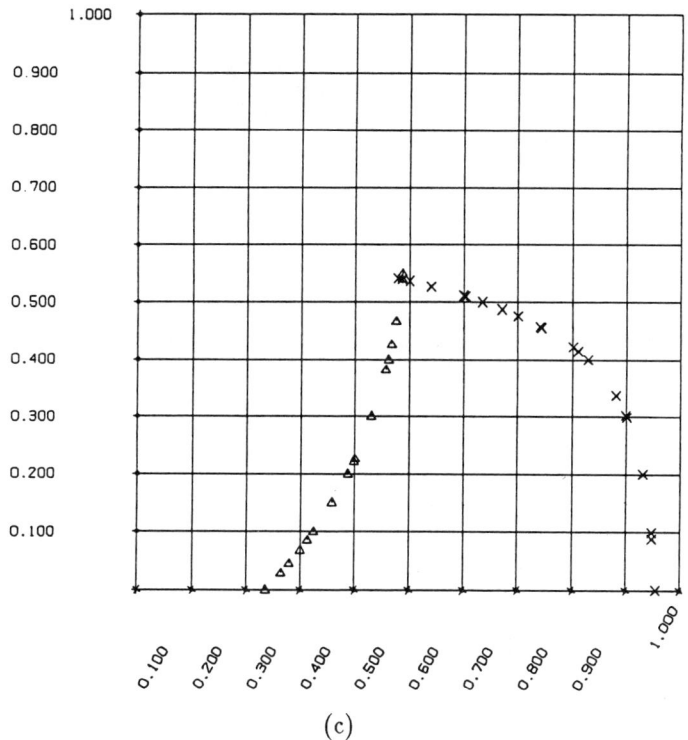

(c)

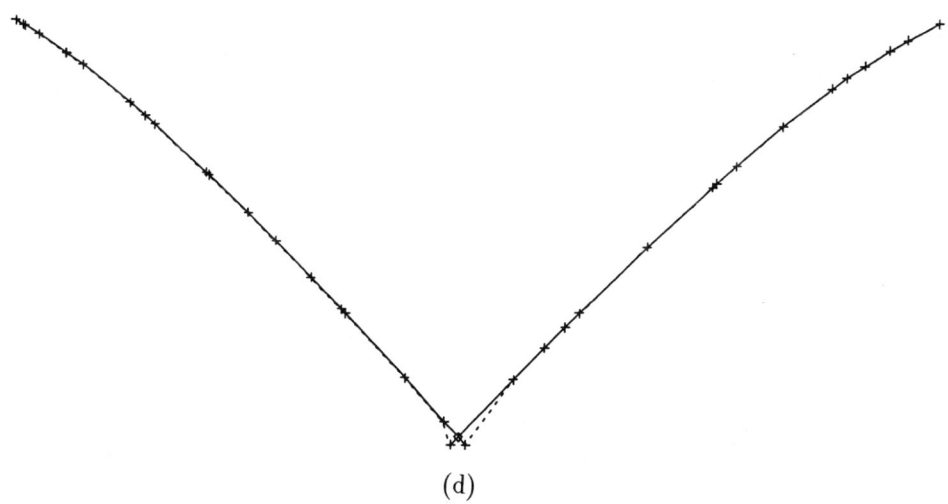

(d)

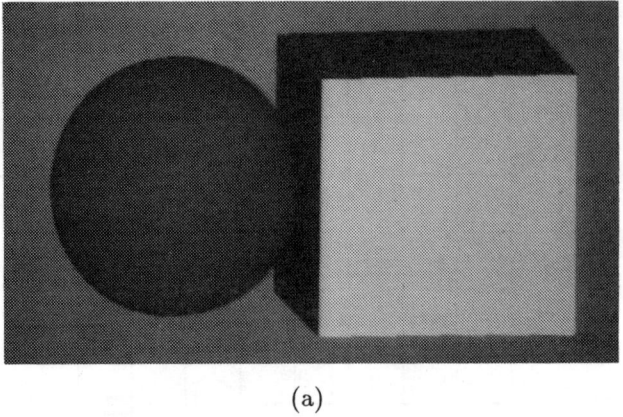

(a)

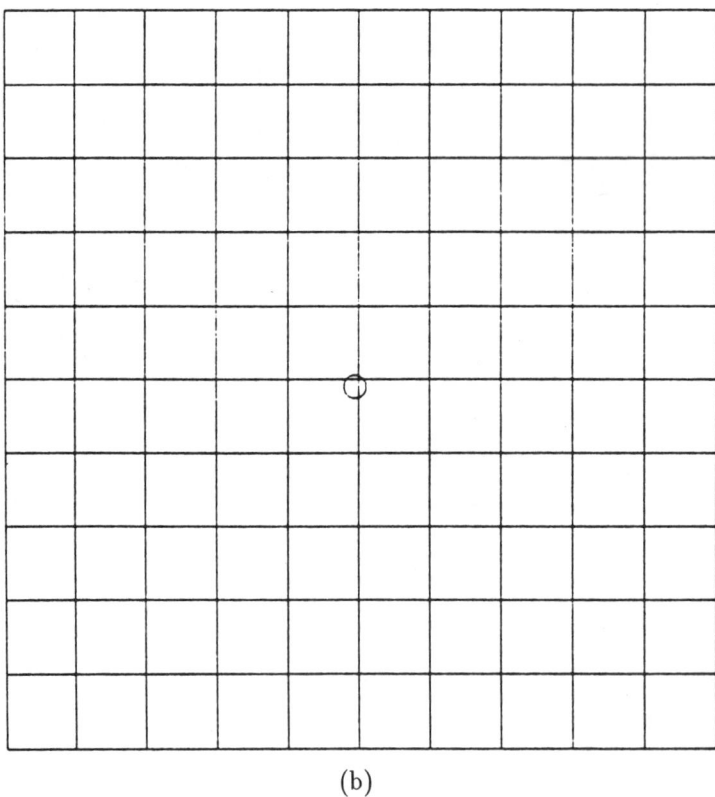

(b)

FIG. 9.8. *Intersections between two geometries that have contact in a very small region:* (a) *a sphere and a cube;* (b) *the intersection curve on one side of the cube.*

between one surface and the isoparametric curves from the other surface. An iterative approach is used for such searching, as it is found to be quite effective. The adaptive control is used to generate more intersection points only when ambiguity in sorting arises or when higher accuracy is needed, and therefore helps reduce the computing time. The coincidence between the two is computed beforehand to separate the case from the iterative search. The cases of tangential surface contact and small loop intersection, which are difficult to evaluate otherwise, can also be identified with a separate minimum distance check. The chances of missing a critical portion of the intersection curves, such as points near tangential contact or branches near a singular point, have been reduced by carrying out the computations on the original geometries, without making approximations.

It should be pointed out that, just as for any other general approach, the method may not be as efficient as the specially designed algorithms when dealing with simple cases. In addition, since the intersections are presented in the form of fitted curves, there will be situations where splines are used to represent specially classified curves, such as circular arcs, even though the specified tolerance is still retained. This less-than-desirable situation can be improved by adding a detection logic before curve fitting starts. For example, by recognizing that circular intersection arcs can result only from a handful of cases (conics intersecting planes, spheres intersecting spheres, etc.), efforts can be made to check if the calculated sampling points fall on a planar circle when surface conditions are met, and then to modify the fitting strategy accordingly.

References

[1] R. E. Barnhill, G. Farin, M. Jordan, and B. R. Piper, *Surface/surface intersection*, Comput. Aided Geom. Des., 4 (1987), pp. 3–16.

[2] J. H. Chuang and C. M. Hoffmann, *On local implicit approximation and its applications*, Technical Report CSD-TR-812, Purdue University, West Lafayette, IN, 1988.

[3] G. A. Crocker and W. F. Reinke, *Boundary evaluation of non-convex primitives to produce parametric trimmed surfaces*, Comput. Graphics, 21 (1987), pp. 129–136.

[4] R. T. Farouki, *The characterization of parametric surface sections*, Computer Vision, Graphics, and Image Processing, 33 (1986), pp. 209–236.

[5] S. L. Hanna, J. F. Abel, and D. P. Greenberg, *Intersection of parametric surfaces by means of lookup tables*, IEEE Comput. Graphics Appl., 3 (1983), pp. 39–48.

[6] E. G. Houghton, R. F. Emnett, J. D. Factor, and C. L. Sabharwal, *Implementation of a divide-and-conquer method for intersection of parametric surfaces*, Comput. Aided Geom. Des., 2 (1985), pp. 173–183.

[7] S. Katz and T. W. Sederberg, *Genus of the intersection curve of two rational surface patches*, Comput. Aided Geom. Des., 5 (1988), pp. 253–258.

[8] R. P. Markot and R. L. Magedson, *Solutions of tangential surface and curve intersections*, Comput. Aided Des., 21 (1989), pp. 421–429.

[9] A. A. G. Requicha and H. B. Voelcker, *Solid modeling: Current status and research directions*, IEEE Comput. Graphics Appl., 3 (1983), pp. 25–38.

[10] T. W. Sederberg, D. C. Anderson, and R. N. Goldman, *Implicit representation of parametric curves and surfaces*, Computer Vision, Graphics, and Image Processing, 28 (1984), pp. 72–84.

[11] P. V. Prakash, *Computation of surface-surface intersections for geometric modeling*, Ph.D. Dissertation, Massachusetts Insitute of Technology, Cambridge, MA, 1988.

[12] S. W. Thomas, *Modelling volumes bounded by B-spline surfaces*, Ph.D. Dissertation, University of Utah, Salt Lake City, UT, 1984.

[13] J. U. Turner, *Accurate solid modeling using polyhedral approximations*, IEEE Comput. Graphics Appl., 8 (1988), pp. 14–28.

[14] W. N. Waggenspack, *Parametric curve approximations for surface intersection*, Ph.D. Thesis, Purdue University, West Lafayette, IN, 1987.

An SSI Bibliography

Gerald Farin

S. Aomura and T. Uehara, *Self-intersection of an offset surface*, Comput. Aided Des., 22 (1990), pp. 417–422.

I. Applegarth, I. Bradley, and D. Catley, *Clipping of B-spline surface patches at surface curves*, in The Mathematics of Surfaces III, D. C. Handscomb, ed., Clarendon Press, Oxford, 1989, pp. 229–242.

P. Arner, *Another look at surface/surface intersection*, Ph.D. Thesis, Department of Mathematics, University of Utah, Salt Lake City, UT, 1987.

C. Asteasu, *Intersection of arbitrary surfaces*, Comput. Aided Des., 20 (1988), pp. 533–538.

D. Ayala, *Boolean operations between solids and surfaces by octtrees: models and algorithms*, Comput. Aided Des., 20 (1988), pp. 452–465.

N. Aziz, R. Bata, and S. Bhat, *Bézier surface/surface intersection*, IEEE Comput. Graphics Appl., 10 (1990), pp. 50–58.

C. Bajaj, C. Hoffmann, J. Hopcroft, and R. Lynch, *Tracing surface intersections*, Comput. Aided Geom. Des., 5 (1988), pp. 285–308.

R. Barnhill, *Geometry processing: curvature analysis and surface-surface intersection*, in Mathematical Methods in Computer Aided Geometric Design, T. Lyche and L. Schumaker, eds., Academic Press, New York, 1989, pp. 51–60.

R. Barnhill, G. Farin, M. Jordan, and B. Piper, *Surface/surface intersection*, Comput. Aided Geom. Des., 4 (1987), pp. 3–16.

R. Barnhill and S. Kersey, *A marching method for parametric surface/surface intersection*, Comput. Aided Geom. Des., 7 (1990), pp. 257–280.

W. Carlson, *An algorithm and data structure for 3D object synthesis using surface patch intersections*, Comput. Graphics, 16 (1982), pp. 255–263.

M. Casale, *Freeform solid modeling with trimmed patches*, IEEE Comput. Graphics Appl., 7 (1987).

V. Chandru and B. Kochar, *Analytic techniques for geometric intersection problems*, in Geometric Modeling: Algorithms and New Trends, G. Farin, ed., Society for Industrial and Applied Mathematics, Philadelphia, 1987, pp. 305–318.

J. Chen and T. Ozsoy, *Predictor-corrector type of intersection algorithm for C^2 parametric surfaces*, Comput. Aided Des., 20 (1988), pp. 347–352.

T. Dokken, *Finding intersections of B-spline represented geometries using recursive subdivision techniques*, Comput. Aided Geom. Des., 2 (1985), pp. 189–195.

T. Dokken and A. Ytrehus, *Recursive subdivision and iteration in intersections and related problems*, in Mathematical Methods in Computer Aided Geometric Design, T. Lyche and L. Schumaker, eds., Academic Press, New York, 1989, pp. 207–214.

R. Farouki, *Trimmed surface algorithms for the evaluation and interrogation of solid boundary representations*, IBM J. Res. Develop., (1987).

Q. Fu, *The intersection of a bicubic patch and a plane*, Comput. Aided Geom. Des., 7 (1990), pp. 475–488.

T. Garrity and J. Warren, *On computing the intersection of a pair of algebraic surfaces*, Comput. Aided Geom. Des., 6 (1989), pp. 137–153.

A. Geisow, *Surface interrogations*, Ph.D. Thesis, University of East Anglia, Norwich, England, 1983.

R. Goldman, *The method of resolvents: a technique for the implicitization, inversion, and intersection of non-planar, parametric, rational cubic curves*, Comput. Aided Geom. Des., 2 (1985), pp. 237–255.

R. Goldman and T. Sederberg, *Analytic approach to intersection of all piecewise parametric rational curves*, Comput. Aided Des., 19 (1987), pp. 282–292.

R. Goldman, T. Sederberg, and D. Anderson, *Vector elimination: a technique for the implicitization, inversion, and intersection of planar parametric rational polynomial curves*, Comput. Aided Geom. Des., 1 (1984), pp. 327–356.

T. Grandine, *Computing zeros of spline functions*, Comput. Aided Geom. Des., 6 (1989), pp. 129–136.

S. Hanna, J. Abel, and D. Greenberg, *Intersection of parametric surfaces by means of lookup tables*, IEEE Comput. Graphics Appl., 3 (1983), pp. 39–48.

C. Hoffmann, *A dimensionality paradigm for surface interrogations*, Comput. Aided Geom. Des., 7 (1990), pp. 517–532.

E. Houghton, E. Emnett, R. Factor, and L. Sabharwal, *Implementation of a divide-and-conquer-method for the intersection of parametric surfaces*, Comput. Aided Geom. Des., 2 (1985), pp. 173–184.

S. Katz and T. Sederberg, *Genus of the intersection curve of two rational surface patches*, Comput. Aided Geom. Des., 5 (1988), pp. 253–258.

A. Kaufmann, *Parallelization of the subdivision algorithm for intersection of Bézier curves on the fps t20*, in Mathematical Methods in Computer Aided Geometric Design, T. Lyche and L. Schumaker, eds., Academic Press, New York, 1989, pp. 403–412.

G. Kriezis, *Algorithms for natural spline surface intersections*, Ph.D. Thesis, Department of Ocean Engineering, Massachusetts Institute of Technology, Cambridge, MA, 1990.

G. Kriezis, P. Prakash, and N. Patrikalakis, *Method for intersecting algebraic surfaces with rational polynomial patches*, Comput. Aided Des., 22 (1990), pp. 645–654.

D. Lasser, *Intersection of parametric surfaces in the Bernstein-Bézier representation*, Comput. Aided Des., 18 (1986), pp. 186–192.

R. Lee and D. Fredericks, *Intersection of parametric surface patches and a plane*, IEEE Comput. Graphics Appl., 4 (1984), pp. 48–51.

J. Levin, *Mathematical models for determining the intersection of quadric surfaces*, Computer Graphics and Image Processing, 11 (1979), pp. 73–87.

R. Markot and R. Magedson, *Solutions of tangential surface and curve intersections*, Comput. Aided Des., 21 (1989), pp. 421–429.

G. Muellenheim, *Convergence of a surface/surface intersection algorithm*, Comput. Aided Geom. Des., 7 (1990), pp. 415–424.

J. Owen and A. Rockwood, *Intersection of general implicit surfaces*, in Geometric Modeling: Algorithms and New Trends, G. Farin, ed., Society for Industrial and Applied Mathematics, Philadelphia, 1987, pp. 335–345.

N. Patrikalakis, *Interrogation of surface intersections*, in Geometry Processing for Design and Manufacturing, R. Barnhill, ed., Society for Industrial and Applied Mathematics, Philadelphia, 1991, pp. 161–185.

―――, *Shape interrogation*, in Automation in the Design and Manufacture of Large Marine Systems, C. Chrystossomidis, ed., Hemisphere, New York, 1990, pp. 83–104.

Q. Peng, *An algorithm for finding the intersection lines between two B-spline surfaces*, Comput. Aided Des., 16 (1984), pp. 191–196.

M. Pratt and A. Geisow, *Surface/surface intersection problems*, in The Mathematics of Surfaces, J. Gregory, ed., Clarendon Press, Oxford, 1986.

M. Sabin, *General interrogations of parametric surfaces*, Technical Report VTO/MS/150, British Aircraft Corporation, 1968.

―――, *A method for displaying the intersection curve of two quadric surfaces*, Computer J., 19 (1976), pp. 336–338.

―――, *Two basic interrogations of parametric surfaces*, Technical Report VTO/MS/148, British Aircraft Corporation, 1968.

R. Sarraga, *Algebraic methods for the intersections of quadric surface patches in GMSOLID*, Comput. Vision, Graphics, and Image Processing, 22 (1983), pp. 222–238.

T. Sederberg, *Algorithm for algebraic curve intersection*, Comput. Aided Des., 21 (1989), pp. 547–556.

T. Sederberg, H. Christiansen, and S. Katz, *Improved test for closed loops in surface intersections*, Comput. Aided Des., 21 (1989), pp. 505–508.

T. Sederberg and R. Meyers, *Loop detection in surface patch intersections*, Comput. Aided Geom. Des., 5 (1988), pp. 161–171.

T. Sederberg and T. Nishita, *Curve intersection using Bézier clipping*, Comput. Aided Des., 22 (1990), pp. 538–549.

T. Sederberg and S. Parry, *A comparison of three curve intersection algorithms*, Comput. Aided Des., 18 (1986), pp. 58–63.

Index